Environmental Contamination Remediation and Management

There are many global environmental issues that are directly related to varying levels of contamination from both inorganic and organic contaminants. These affect the quality of drinking water, food, soil, aquatic ecosystems, urban systems, agricultural systems and natural habitats. This has led to the development of assessment methods and remediation strategies to identify, reduce, remove or contain contaminant loadings from these systems using various natural or engineered technologies. In most cases, these strategies utilize interdisciplinary approaches that rely on chemistry, ecology, toxicology, hydrology, modeling and engineering.

This book series provides an outlet to summarize environmental contamination related topics that provide a path forward in understanding the current state and mitigation, both regionally and globally.

Topic areas may include, but are not limited to, Environmental Fate and Effects, Environmental Effects Monitoring, Water Re-use, Waste Management, Food Safety, Ecological Restoration, Remediation of Contaminated Sites, Analytical Methodology, and Climate Change.

C. D. Metcalfe · Erin R. Bennett
Editors

Building Resilience to Climate Change in Small Island Developing States (SIDS) in the Caribbean

Springer

Editors
C. D. Metcalfe
School of the Environment
Trent University
Peterborough, ON, Canada

Erin R. Bennett
School of the Environment
Trent University
Peterborough, ON, Canada

ISSN 2522-5847 ISSN 2522-5855 (electronic)
Environmental Contamination Remediation and Management
ISBN 978-3-031-37375-6 ISBN 978-3-031-37376-3 (eBook)
https://doi.org/10.1007/978-3-031-37376-3

This Springer imprint is published by the registered company Springer Nature Switzerland AG
The registered company address is: Gewerbestrasse 11, 6330 Cham, Switzerland

Foreword

The vulnerability of the economies, societies, and natural ecosystems of Small Island Developing States (SIDS) in the Caribbean to external shocks is well documented, not only in the literature on sustainable development, climate change, and risk management, but also on the psyche of Caribbean people. This book contains frequent references to a slew of extreme weather events that have severely impacted the region over the past two decades, from Hurricane Ivan in 2004 to Hurricane Irma in 2017.

The consensus from much of the literature on vulnerability is that because social, economic, and ecological systems are intimately linked, negative shifts in any of these systems, caused by internal and/or external shocks, can make countries more vulnerable and can lead to a loss of resilience, thus negatively impacting their sustainable development. It is generally accepted that failure to anticipate shocks can lead to partial solutions that do not take account of the natural integration between social, environmental, and economic processes and ignore the returns from resilient solutions.

The Resilience Alliance defines resilience as applied to integrated systems of people, economies, and nature as (a) the amount of disturbance a system can absorb and still remain within the same state or domain or attraction; (b) the degree to which the system is capable of self-organization (versus lack of organization or organization forced by external factors); and (c) the degree to which the system can build and increase the capacity for learning and adaptation (Carpenter, 2001).[1]

Against this background, all Caribbean Community (CARICOM) Member States are seeking to build their social, economic, and environmental resilience in areas in which they are most vulnerable, such as exposure to energy price shocks; inherent vulnerability to climate change and natural disasters; vulnerability of natural ecosystems to stressors arising from population growth and unsustainable demand for and use of these resources; and exposure to commodity price shocks and foreign exchange and financial crises. Many regional treaties such as the Revised Treaty of

[1] https://groups.nceas.ucsb.edu/sustainability-science/2010%20weekly-sessions/session-102013-11.01.2010-emergent-properties-of-coupled-human-environment-systems/supplemental-readings-from-moderator-discussant-jim-heffernan/Carpenter%20et%20al%202001%20Meta.html.

Chaguaramas (RTC) and the Revised Treaty of Basseterre establishing the Organization of Eastern Caribbean States Economic Union (ECCU) commit the region's governments to adopt effective measures to manage their natural resources, including measures for the effective management of the soil, air, and all water resources, their Exclusive Economic Zones (EEZs), and all other maritime areas under their national jurisdiction; conserve biological diversity; sustainably use biological resources; minimize their environmental vulnerability; improve environmental management; and protect the region's natural (including historical and cultural) resource base for optimal social and economic benefits.

Building resilience to climate variability and climate change has, with good reason, emerged as a top national and regional imperative. While the Caribbean's contribution to greenhouse gas emissions that cause global climate change is negligible, the region has suffered and is projected to suffer even more from intense hurricanes, changes in rainfall patterns, and sea level rise if global warming is not kept under 1.5 °C. The IPCC's Special Report predicts that warming of 1.5 °C will destroy between 70 and 90 percent of reef-building corals, while warming of 2 °C will likely destroy 99 percent of tropical coral reefs (Gergis, 2019).[2]

Of major concern to the region are the pathways through which the impacts of climate variability and climate change can be magnified. Here, two inter-related threats loom large, namely land degradation and urbanization. Land degradation is manifested in accelerated deforestation, soil erosion, a decline in the fertility and productivity of land, deterioration in water quality, and the destruction of ecosystems in lagoons. Urbanization, when combined with inherent challenges of smallness and social, economic, and environmental vulnerability, presents serious challenges for Caribbean Governments. The visual evidence in most cities in the Caribbean reflects unplanned or inadequately managed expansion and urban sprawl. These drivers, in turn, contribute to other challenges, such as land degradation, congestion, dysfunction, inadequate waste disposal, air and water pollution, and increased exposure to natural hazards. Consequently, managing urbanization requires an integrated approach to addressing the challenges and opportunities associated with inclusive, resilient, and sustainable development.

The essential appeal of this book is that it treats these longstanding and emerging barriers to building resilience in Caribbean SIDS in an integrated manner. In this respect, planners and policy makers will find value in the policy ideas, insights, and practical solutions that are identified, many of which are being applied within the region and across the SIDS community. Critically, this book emphasizes the need for the region to establish a resilience culture that is rooted in citizens, businesses, and institutional resilience. It correctly notes that the absence or weakness of one can place an inordinate strain on the others. The emphasis on business resilience is especially noteworthy, as the private sector is not often included in the dialogue on resilience building.

---

[2] https://www.themonthly.com.au/issue/2019/august/1566136800/jo-lle-gergis/terrible-truth-climate-change.

I am delighted not only with the content of this book but also with the fact that it emerged from a meeting of minds from the Caribbean and around the world, driven by an innate desire to contribute to the sustainability and resilience of the economies, societies, and natural ecosystems of Caribbean SIDS. It is my sincere hope that this book will be marketed and mined in such a way that it becomes a living document in schools, boardrooms, research and innovations labs, and community-based organizations.

Castries, St. Lucia Cletus I. Springer

Cletus I. Springer served for 12 years as Director of the Department of Sustainable Development of the Organization of American States (OAS), based in Washington D.C. He has written extensively on the sustainable development of SIDS. At the time of publication he served as Chairman of the Caribbean Natural Resources Institute (CANARI), and the Sir Arthur Lewis Community College (SALCC) in his native Saint Lucia.

Preface

Small Island Developing States (SIDS) in the Caribbean are among the most vulnerable regions in the world to the effects of climate change, which include increased frequency and intensity of extreme weather events (i.e., hurricanes, droughts), flooding, and sea level rise. For instance, for some Caribbean islands impacted by recent hurricanes, damage to infrastructure and livelihoods has exceeded the size of the economy. The impacts of coastal damage, flooding, and droughts in the Caribbean are predicted to worsen during this century. The usual approach to protect against these impacts in Caribbean SIDS is to react with "hard" engineering solutions. These structural engineering approaches prevent damage and disruptions to services associated with predictable events but leave communities vulnerable to future events that do not follow historical trends. Furthermore, engineered structures do not adequately address the systemic nature of climate change nor account for compounding threats (e.g., coincidence of hurricane season and global pandemics).

To move from this traditional strategy for managing risks from these hazards, we need to consider a portfolio of solutions that enhance island protection and community resilience. For instance, Nature-based Solutions (NbS) are gaining attention as practical and cost-effective approaches for mitigating climate-based stressors. However, deployment of NbS strategies requires spatial coordination within the context of "ridge to reef" or integrated water resource management (IWRM) approaches that include the creation of conditions for social acceptance, equity, effective governance, and financial incentives. Approaches that integrate the environmental, economic, and physical domains with the values and needs of the population are necessary to develop sustainable strategies that will enhance the resilience of small islands, within the context of inter-island differences in geology, ecology, societal attitudes, governance, and human and economic resources.

The chapters in this book summarize some of the concepts discussed during a three-day workshop held in Barbados in June 2022 which brought together experts from the Caribbean territories of the USA and from CARICOM countries, as well as colleagues from the USA, Canada, Europe, and the U.K. to share their experiences in building resilience to protect coastal zones and inland areas from the impacts of extreme weather events and associated cascading hazards. The discussions focused on strategies for building capacity among Caribbean countries to build resilience to climate change. There are currently few examples of the successful application of these strategies in the Caribbean. Where they do exist, scaling up lessons and approaches have been limited. The workshop provided a forum for colleagues from the Caribbean to share their experiences and to present case studies of NbS and community-based programs for building resilience, as well as to discuss challenges and solutions for overcoming barriers to action. The chapters in this book provide summaries of some of these case studies from colleagues working in Trinidad and Tobago, the U.S. Virgin Islands, St. Lucia, Grenada, and St. Kitts and Nevis.

These discussions also were relevant to several of the UN Sustainable Development Goals, including SDG 14 (Life Under Water), SDG 15 (Life on Land), SDG 13 (Climate Action), and SDG 6 (Clean Water and Sanitation), as well as the UN Sendai Goals for Disaster Risk Reduction. The workshop was also relevant to both the Paris Agreement goals to combat climate change and the Cartagena Convention for the protection of the Caribbean Sea. The workshop also explored the synergies that can be achieved by applying the concept of the "water-energy-food-climate nexus". This book not only makes a valuable contribution to the literature on building resilience within the context of small islands in the Caribbean, but also has relevance for other island nations around the world that must address the challenges of climate change.

C. D. Metcalfe
Professor Emeritus
Trent University
Peterborough, ON, Canada

Erin R. Bennett
Adjunct Professor
Trent University
Peterborough, ON, Canada

Contents

Contributors

Erin R. Bennett The School of the Environment, Trent University, Peterborough, ON, Canada

Adrian Cashman AKWATIX Water Resource Management, St. John, Barbados

Christopher Corbin The Caribbean Environment Programme, United Nations Environment Programme, Kingston, Jamaica

Shelly-Ann Cox Barbados Fisheries Division, Ministry of Environment and National Beautification, Green and Blue Economy, St. Michael, Barbados

Christopher Cummings Engineer Research and Development Center, US Army Corps of Engineers, Vicksburg, MS, USA

Nikolai Emmanuel Vetiver Network International, Port of Spain, Trinidad and Tobago

Nealla Frederick The Nature Conservancy Caribbean Programme, St. George's, Grenada

Stephanie E. Galaitsi Engineer Research and Development Center, US Army Corps of Engineers, Vicksburg, MS, USA

Gregory E. Guannel Caribbean Green Technology Centre, University of the Virgin Islands, St. Thomas, Virgin Islands (U.S.)

Genora Joseph The Climate Resilience Agency for Dominica (CREAD), Roseau, Dominica

Jesse Keenan School of Architecture, Tulane University, New Orleans, LA, USA

Gregory A. Kiker Department of Agricultural and Biological Engineering, University of Florida, Gainesville, FL, USA

James H. Lambert Centre for Risk Management of Engineering Systems, University of Virginia, Charlottesville, VA, USA

Igor Linkov Engineer Research and Development Center, US Army Corps of Engineers, Vicksburg, MS, USA

Patrick McConney Center for Resource Management and Environmental Studies, University of the West Indies, St. Michael, Barbados

C. D. Metcalfe The School of the Environment, Trent University, Peterborough, ON, Canada

Jose M. Palma-Oliveira Faculty of Psychology, CIPSI, University of Lisbon, Lisbon, Portugal

Gianluca Pescaroli Institute for Risk and Disaster Reduction, University College London, London, UK

Rebecca A. Rebar Centre for Risk Management of Engineering Systems, University of Virginia, Charlottesville, VA, USA

Halla R. Sahely Independent Environmental Consultant, Basseterre, St. Kitts and Nevis

Chante Saunders Center for Resource Management and Environmental Studies, University of West Indies, St. Michael, Barbados

Cletus Springer Caribbean Natural Resources Institute (CANARI), Sir Arthur Lewis Community College (SALCC), Castries, Saint Lucia

Roxann K. Stennett-Brown Department of Engineering, University of West Indies, Mona Campus, Kingston, Jamaica

Burton C. Suedel US Army Engineer Research and Development Center, Vicksburg, MS, USA

Benjamin D. Trump Engineer Research and Development Center, US Army Corps of Engineers, Vicksburg, MS, USA

Kim Waddell Caribbean Green Technology Centre, University of the Virgin Islands, St. Thomas, US Virgin Islands

Chapter 1
An Introduction to Small Island Developing States and Building Resilience to Climate Change in the Caribbean

C. D. Metcalfe, Erin R. Bennett, and Kim Waddell

Abstract This article describes the unique features and characteristics of Small Island Developing States (SIDS) in the Caribbean that contribute to their vulnerability to the impacts of climate change and describes the regional barriers to building resilience to mitigate these impacts. As an illustration of the efforts needed to overcome these barriers, examples from the US Virgin Islands are described on how to build technical capacity, manage monitoring information, foster collaborations among partners, communicate risks to communities, and sustainably manage natural resources.

1.1 Introduction

In an online report published by the International Monetary Fund (IMF), Muñoz and Ötker (2018) made the case that nations in the Caribbean are among the most vulnerable regions in the world to the effects of climate change, with increased frequencies and intensities of hurricanes being the most devastating example. For some extreme weather events in the Caribbean, damages exceed the size of the economy. For example, the costs of damage to the island of Dominica caused by Hurricane Maria in 2017 amounted to 226% of annual GDP. Climate change is expected to intensify these vulnerabilities in the Caribbean, as the warming global climate fuels

C. D. Metcalfe (✉) · E. R. Bennett
The School of the Environment, Trent University, Peterborough, ON, Canada
e-mail: cmetcalfe@trentu.ca

E. R. Bennett
e-mail: erbennett@trentu.ca

K. Waddell
Caribbean Green Technology Centre, University of the Virgin Islands, St. Thomas, US Virgin Islands
e-mail: kim.waddell@uvi.edu

C. D. Metcalfe and E. R. Bennett (eds.), *Building Resilience to Climate Change in Small Island Developing States (SIDS) in the Caribbean*, Environmental Contamination Remediation and Management, https://doi.org/10.1007/978-3-031-37376-3_1

more frequent and intense storms and contributes to rising sea levels (Reyer et al. 2017). Two geographic features make SIDS highly susceptible to extreme weather events and to sea level rise: (i) they have long coastlines relative to land area and (ii) a high proportion of their coastal lands are low lying, and this is where much of the population lives. Many SIDS are therefore susceptible to inundation of coastal zones, beach and shoreline erosion, and saltwater intrusion into aquifers (Cambers et al. 2008). Also, because of the small land areas and often rugged inland topography of Caribbean SIDS, "retreat strategies" to other less vulnerable areas of the island are not an option for the island populations (Mycoo 2014).

Muñoz and Ötker (2018) identified several strategies for building resilience to disasters and climate risks in the Caribbean, including "*structural protection to reduce disaster risks, such as through resilient infrastructure*". These authors also advised that island nations will need to implement and enforce adequate land-use management practices, zoning rules, and building codes, as well as develop social safety nets. This chapter provides a review of both the barriers and the opportunities for building resilience to natural disasters and sea level rise in Caribbean SIDS.

1.2 Context for Building Resilience in Caribbean SIDS

The Greater Caribbean Region (GCR) includes many islands, as well as the countries bordering the Caribbean Sea on the mainland of South America and Central America, and the eastern coast of the Yucatan Peninsula in Mexico (Fig. 1.1). The islands in the Bahamas and the Turks and Caicos, while sometimes considered part of the GCR, are actually located in the western Atlantic. The island nations within the Caribbean Sea vary in size, economies, systems of governance, language, and culture. Therefore, some context is required to identify the SIDS within the Caribbean. First, "small islands" are defined by the United Nations Educational, Scientific, and Cultural Organization (UNESCO) as islands with areas less than 2,000 km^2 or widths less than 10 km. This definition excludes the larger Caribbean islands of Cuba, Hispaniola (Dominican Republic and Haiti), and Jamaica. Medium-sized islands such as Puerto Rico (9,100 km^2) and Trinidad (5,100 km^2) are also vulnerable to extreme weather events, as illustrated by the destruction in Puerto Rico caused by Hurricane Maria in 2017. Several island nations in the Caribbean are an archipelago of smaller islands; some of which are settled by humans and some not. For instance, about 50 islands constitute the U.S. Virgin Islands, with the three largest being St. Croix, St. Thomas, and St. John. There are 35 islands included in the Grenadines group that are located between the larger islands of Grenada and St. Vincent.

Second, there is no consistent definition to identify countries as "developing". The United Nations (UN) classifies countries as either developed economies, economies in transition, or developing economies based on general criteria that "reflect basic economic country conditions". The World Bank uses gross income per capita for its measurements and, using these criteria, classifies a high percentage of Caribbean nations as "high-income" economies, meaning they have a gross per capita income

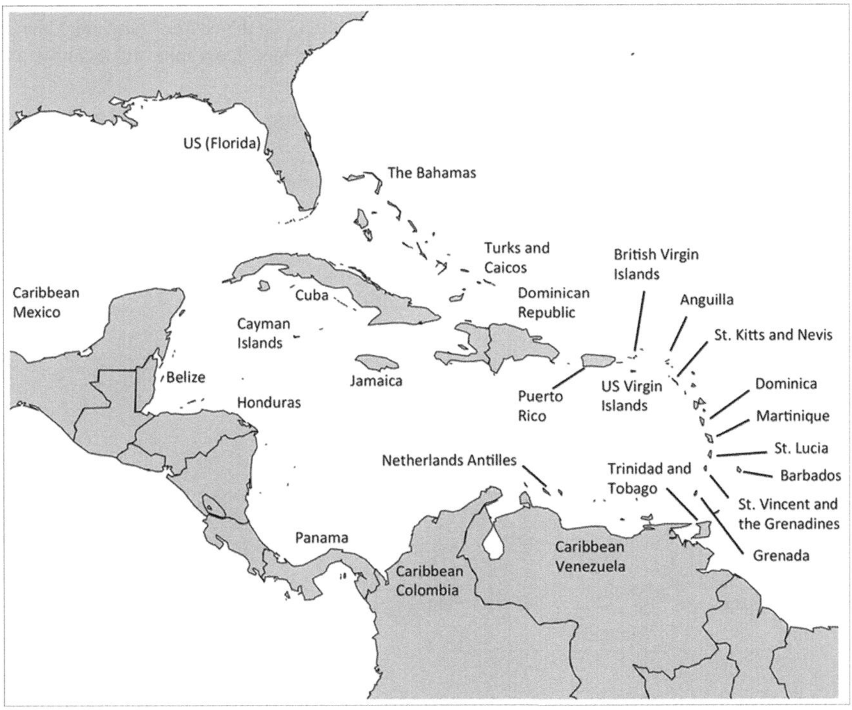

Fig. 1.1 Map of the Greater Caribbean Region (GCR), also including the Bahamas and the Turks and Caicos

that exceeds US$12,475 (World Bank 2016). These countries include Antigua and Barbuda, Barbados, Puerto Rico, Sint Maarten, St. Kitts and Nevis, Trinidad and Tobago, the British Virgin Islands, and the US Virgin Islands. All other Caribbean countries, except Haiti, are classified by the World Bank as "upper-middle income" economies. The human development index (HDI) was developed by the UN as a metric to assess levels of social and economic development. HDI quantifies life expectancy, educational attainment, and income into a standardized number between zero and one; the closer to one, the more developed the country. By application of these metrics, the Cayman Islands is often disqualified from being included within Caribbean SIDS because of the HDI of 0.888. However, several other Caribbean islands have an HDI exceeding 0.8 (Table 1.1).

There are several island nations located within the Caribbean Sea that meet one or more of the criteria for identifying them as SIDS (Table 1.1). It must be recognized, however, that the classification used here differs from the UN Office of the High Representative for the Least Developed Countries, Landlocked Developing Countries, and Small Island Developing States that regard most of the islands in the GCR plus the Bahamas, as well as some countries on the mainland (i.e., Guyana, Suriname, Belize) as Caribbean SIDS (UN-OHRLLS 2011). As summarized in Table 1.1, the

Table 1.1 Islands in the Caribbean that can be considered SIDS on the basis of land area, human development index (HDI), or size of the population. Islands are listed geographically from north to south in the Caribbean Sea

Island group	Islands	Area (km^2)	Geology	Population	HDI
Puerto Rico		9,100[a]	Volcanic and marine sediments	3.2 million	0.845
US Virgin Islands	St. Croix	218	Volcanic with limestone and conglomerates	89,000	0.894
	St. Thomas	83			
	St. John	50			
British Virgin Islands	Tortola	56	Volcanic with limestone and conglomerates	36,000	–
	Virgin Gorda	8			
Saba		13	Volcanic	1,900	–
St. Eustatius		21	Volcanic	3,100	–
Anguilla		102	Limestone capping volcanic rock	15,100	–
St. Maarten/St. Martin[b]	St. Maarten	34	Volcanic	42,900	0.702
	St. Martin	96		40,000	–
St. Barthélemy		21	Volcanic	9,130	–
St. Kitts and Nevis	St. Kitts	261	Volcanic and sedimentary	53,200	0.779
	Nevis	93		11,100	
Antigua and Barbuda	Antigua	440	Limestone capping volcanic rock	99,900	0.778
	Barbuda	161			
Montserrat		102	Volcanic	4,990	0.821
Guadeloupe		1,628	Volcanic	396,000	–
Dominica		750	Volcanic	72,000	0.742
Martinique		1,128	Volcanic	376,500	–
St. Lucia		617	Volcanic	183,600	0.759
St. Vincent and the Grenadines		389	Volcanic	110,900	0.738
Barbados		430	Karst capping sedimentary rock	287,400	0.814
Grenada		344	Volcanic	112,500	0.779
Trinidad and Tobago	Trinidad	5,128[a]	Conglomerates, sandstone, lavas, and limestone	1.4 million	0.796
	Tobago	300			
Netherland Antilles (Leeward Islands)	Aruba	180	Conglomerates and limestone, overlaying lava	201,000	–
	Bonaire	288		20,100	

(continued)

Table 1.1 (continued)

Island group	Islands	Area (km^2)	Geology	Population	HDI
	Curacao	444		155,000	

[a]Medium-sized islands by the UNESCO definition
[b]St. Maarten (Netherlands) and St. Martin (France) share a border dividing the same island

land areas, geology, population sizes, and HDI of Caribbean SIDS vary widely, but there are some common features that contribute to challenges for building resilience to natural disasters and sea level rise. There are also differences among the island nations in their vulnerability that have an influence on the strategies required to build resilience.

1.3 Barriers to Building Resilience Within Caribbean SIDS

The results of a survey of senior policy makers from several Caribbean SIDS indicated that the two most cited barriers to building resilience to climate change were insufficient finances and a lack of technical capacity (Robinson 2018). Recent emigration patterns for the Caribbean show migration of island residents primarily for economic and educational purposes to metropolitan centers in the USA, Canada, and the UK (Thomas and Benjamin 2018). This has led in many cases to a shortage of trained professionals with the qualifications to address the environmental and engineering challenges posed by climate change (Mercer et al. 2012). Research capacity is also limited across the Caribbean, as investment in research is less than ½% of GDP (Robinson 2018). Lack of financial resources in Caribbean SIDS means that there is a reliance on international sources of funding to finance programs to build resilience to climate change and sea level rise (Robinson 2018). Since these programs are generally project-based and have limited timescales (e.g., 4–5 years), it is a challenge to fully implement projects that are slow to develop, such as nature-based solutions (NbS). In addition, the bidding process for projects funded by international agencies may be biased toward large engineering firms from outside the Caribbean that favor "gray" or engineered infrastructure because it is a more widely accepted approach and can be implemented over shorter timescales.

The reliance on the tourism sector for revenue among many Caribbean SIDS may also be a barrier to implementing projects that build resilience (Mycoo 2014). In a study conducted in Barbados, Weaver (2001) reported that unsustainable land management practices favoring mass tourism and high-density beach accommodation were major factors in the destruction of fragile ecosystems. For instance, only a small area of coastal mangrove remains in Barbados because of hotel development. Therefore, the tourism sector may not be supportive of approaches that reduce beach access, such as projects to restore or protect mangroves. However, studies that show that tourists will avoid returning to destinations with degraded ecosystems, beach erosion, and damaged coral reefs (Uyarra et al. 2005) may be changing attitudes within the tourism sector and encouraging decision-makers to develop and enforce more sustainable land management practices (Mycoo 2014). The development of comprehensive land-use management practices, rules of zoning and stringent building codes, and the enforcement of these regulations will be key to building resilience (Muñoz and Ötker (2018).

However, systems of governance vary widely among Caribbean islands and there are varying capacities among the nations and communities for cooperation, raising revenues, decision making, and long-range planning. For instance, St. Vincent and the Grenadines do not have a provincial government, and administrative functions are carried out by six independent parishes. Politics in Grenada takes place in a framework of a parliamentary democracy, with a representative of the British Crown as the Head of State. Sint Maarten, a territory of the Netherlands, and Saint Martin, a territory of France, share the same island, but have separate elected governments and heads of state. The governance of the US Virgin Islands is within a framework of a "democratic dependency" of the USA, whereby the elected Governor is the head of the local government and party leader in a multi-party system. These many systems of government and, in some cases, economic and governance ties to other nations (e.g., U.K., USA, the Netherlands, France) create organizational challenges for building resilience in Caribbean SIDS. Regional alliances, such as the CARICOM political union, may help to overcome some of these barriers (CCCCC 2012).

The SIDS in the region vary in their vulnerability to extreme weather events because of their geographic location within the Caribbean Sea. Although the entire GCR is subject to tropical storms, the intensity and frequency of extreme weather events vary throughout the region. For instance, the last storm that caused any significant damage or loss of life in Trinidad was a 1933 hurricane. Hurricane Flora in 1963 (category 3) was the last storm to cause deaths and significant damage on the island of Tobago. Because of its easterly location, Barbados is rarely hit by hurricanes, although in July of 2021, Hurricane Elsa passed just south of the island as a category 1 storm. In contrast, the US Virgin Islands, which are located within the "hurricane belt" of the Caribbean, have been severely impacted in recent years by multiple extreme storm events, including Hurricane Earl (category 1) in 2010, Hurricane Irma and Hurricane Maria (both category 5) within weeks of each other in 2017, and Hurricane Dorian (category 1) in 2019. However, Caribbean SIDS also differ in degrees of preparedness for these extreme events (Rhiney 2015).

Differences in geology among Caribbean SIDS are factors affecting the vulnerability to natural disasters. Over a distance of 700 km, from the island of Anguilla in the north to Grenada in the south, volcanic islands form an arc along the eastern border of the Caribbean (Table 1.1). These islands, several of which are still volcanically active (i.e., St. Vincent, Martinique, Guadeloupe, and Montserrat) typically have steep topographies and relatively impermeable substrates, contributing to flooding during extreme weather events. For instance, in 2010, the Soufríere River watershed on the rugged southwest coast of St. Lucia was heavily impacted by Hurricane Tomás (Category 2), and flooding and landslides caused major damage to infrastructure, as well as loss of life. The surface substrate of some other SIDS is limestone formed by coral capping of volcanic seamounts. Barbados is unusual among Caribbean islands in having a karst substrate formed by coral capping of ancient sedimentary rock forced to the sea surface. These differences in geology among Caribbean SIDS influence the hydrological regimes and the types of interventions that can be employed to build resilience.

1.4 Overcoming Barriers to Building Resilience

As mentioned previously, insufficient finances and a lack of technical capacity are the two most cited barriers to building resilience to climate change identified by senior policy makers in the Caribbean (Robinson 2018). However, there are exciting new initiatives underway in the region to overcome these barriers. These include the Green Climate Fund to address the need for long-term financing of projects for building resilience to climate change, and programs to enhance training in the Caribbean in the fields of Science, Technology, Engineering, and Mathematics (STEM) and as data scientists. As an example of efforts in the region to build technical capacity for building resilience, what follows is a description of the ongoing effort to build capacity in the US Virgin Islands (USVI); specifically, within the public sector of the USVI, which is the largest employer in the Territory to meet the challenge of climate change and build resilience to the impacts of natural hazards.

Building technical capacity to meet the needs of the USVI is a challenge that the University of the Virgin Islands (UVI) has taken on since its inception in 1964 as the only institution of higher learning in the Territory. UVI researchers, in partnership with the Virgin Islands Territorial Emergency Management Agency (VITEMA), are developing the next Hazard Mitigation and Resilience Plan (HMRP) for the Territory. The primary goal of hazard mitigation planning is to reduce loss of life and property by minimizing the impact of disasters. Such plans, typically developed by state or local governments, initially identify natural disaster risks and vulnerabilities that are locally common. After identifying these risks, the plans recommend long-term strategies for protecting people and property (FEMA 2022). The USVI is part of Region 2 of the Federal Emergency Management Agency (FEMA) of the USA that includes New York, New Jersey, Puerto Rico (PR), and the USVI. Hurricane Sandy, one of the costliest storms (i.e., nearly US$ 70 billion in damages) in U.S. history at the time (Diakakis et al. 2015), impacted the northern Region 2 states of New York and New Jersey. FEMA's mitigation planners based in those states learned a great deal from that storm and applied those lessons when developing new strategies for mitigation, response, and recovery efforts; especially following the 2017 hurricanes Irma and Maria that devastated PR and the USVI (FEMA planner John Heide, pers. comm.). One of those lessons was the realization that resilience planning that integrates existing and projected climate change impacts, coupled with implementation, has significant potential to reduce costs in lives and property.

Resilience planning works to integrate resilience into community planning and land use. It also goes beyond the traditional hazard mitigation strategies that commonly focus on the built environment (i.e., critical infrastructure and property in both the public and private sectors) by facilitating a community's ability to prepare and plan for, as well as absorb and recover from disaster-related or other adverse events. Resilience planning addresses the underlying conditions that are barriers to the rapid recovery of communities and their systems that provide modern services such as power, water, telecommunication, and transportation following these events (https://resilientvi.org/).

In addition to integrating principles of **resilience planning** and **climate change** impacts, the Virgin Islands' HMRP advocates for **sustainable development** by harmonizing resource use, directed investments, and technological advances that can meet the current and future societal needs with diminished ecosystems and likely population growth, as recognized by the Brundtland Commission in 1987. However, arguably the most relevant principle for discussion in this chapter is that of **Capacity Building;** that is, the process by which people, organizations, and society systematically stimulate and develop their capability over time to achieve social and economic goals, including through improvement of knowledge, skills, systems, and institutions within a wider social and cultural enabling environment. This may be the greatest challenge facing SIDS in the Caribbean with regards to building resilience against the impacts of climate change.

Caribbean states and territories like the USVI struggle to implement, monitor, and evaluate progress in the achievement of national or international sustainable development goals for a variety of reasons. It is difficult to develop and follow indicators of progress without:

- environmental regulations supported and sustained consistently by local government,
- coordination among government agencies that produce statistics and other relevant data in a timely fashion,
- allocation of resources to sustain essential technical staff needed to support monitoring, regulatory enforcement, and analyses (ECLAC 2019). These are many of the same challenges facing Caribbean SIDS as they attempt to enhance resilience.

As noted earlier, the HMRP for the USVI addresses vulnerabilities to natural hazards including climate change by defining approaches that enhance the capabilities of individuals, communities, government, institutions, and businesses to withstand, survive, adapt, and recover from disruptions that may not be predictable, such as the two Category 5 hurricanes that arrived two weeks apart in the Northeastern Caribbean in 2017, or the global pandemic that began in 2020.

Figure 1.2 maps out the processes and outcomes specific to UVI's approach to creating the HMRP. This approach integrates our core principles of resilience and sustainability into the Process. It also includes the multi-sectoral "Stakeholder Engagement" effort and the "Assessment and Analyses" of hazards and risks. The

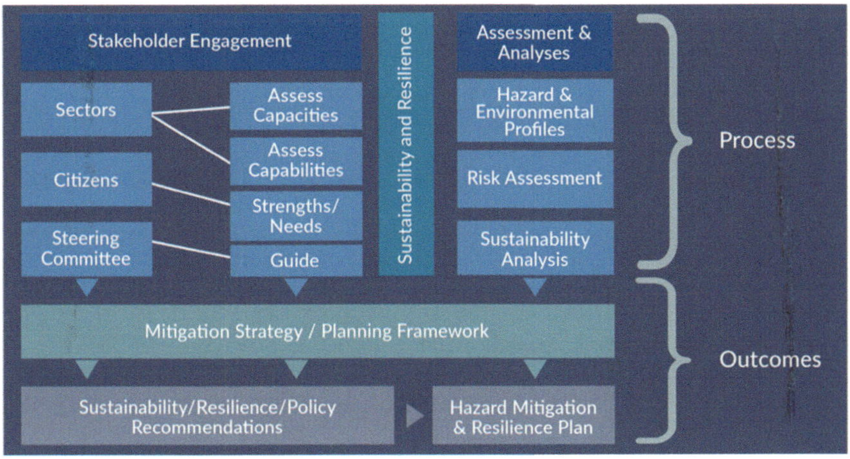

Fig. 1.2 US Virgin Islands Hazard Mitigation and Resilience Plan (HMRP) illustrating the framework and the approach. Adapted from Resilientvi.org (2022)

products of these discussions and analyses then inform Outcomes such as the Mitigation Strategies and Planning Framework, policy recommendations, and eventually the final outcome—the Hazard Mitigation and Resilience Plan.

Through the assessment and analysis part of the process, the UVI team identified a nuanced but important difference between strategies for mitigating hazards in the future. Most hazard mitigation plans, much like sustainable development plans, focus on the built environment by repairing or rebuilding the "robustness" of their systems. Robustness is the ability or capacity of a system to resist or to take damage while retaining its normal processes to at least some degree. Meanwhile, resilience is the recovery that occurs within a system after some form of disturbance to the system.

The USVI qualifies and receives significant financial support from the U.S. federal government following disasters, but this support routinely focuses on re-establishing the robustness of the damaged systems and infrastructure following disasters. However, the capacity of local government to manage the funding or the rebuilding projects has been shown to overwhelm local managers (HMRP, unpublished data). Coupled with federally required funding protocols and procedures that are not typically sensitive to island history, practices, or territorial capacity, much of this funding often remains unspent. The HMRP has developed recommendations that focus on enhancing both the robustness and the resilience of these systems (Table 1.2) but emphasizes the importance of integrating resilience into planning efforts that address natural hazards and climate change.

If Caribbean states and territories hope to increase resilience in the face of climate change, they need to build capacity. This can be prioritized by targeting three guiding objectives. The first is "**Think Systems**" where public and private sectors together articulate the importance of data-informed decision making with a goal of providing timely services to the islands' populations. Government agencies must break down

Table 1.2 Conceptual Cost–Benefit-Analysis contrasting Robustness and Resilience strategies in the U.S. Virgin Islands

	Robustness	Resilience
Projects	• Rebuild damaged infrastructure, • Invest in efforts to resist the impacts of the hazard, • Modify the built environment	• Focus on operations and adequate staffing, • Support Public/Private Partnerships and Coordination, • Implement comprehensive land and water use plans, • Maintain existing infrastructure and services
Costs	• Funding from Federal Grants and other sources, • Need contractors and other outside workforce	• Usually not covered by federal support, • Investment in people training and management, • Time-sensitive
Benefits	• Ability to withstand and resist Impacts, • Timely restoration of services after expected disruptions	• Survive surprises, • Anticipate and adapt to both expected and unexpected, • Stronger integration of activities and sharing of knowledge among different communities, • Build shared identity and values among the population

their mission-driven silos and recognize their reliance on each other to effectively deliver the services when needed. Finally, technologic advances can simplify and streamline operations within all island enterprises through training and adoption of new tools and methods.

Addressing climate change impacts is fundamentally about **managing risk**. This is the second of our guiding objectives. Caribbean governments and communities need to understand the hazards that they are likely to face, but also anticipate surprises. To do this requires investments in mitigation strategies. For financially strapped island nations that might be reluctant to spend on mitigation, it may be persuasive to know that FEMA and the National Institute of Building Sciences estimated that for every dollar spent on mitigation saved US$6 in disaster recovery costs (FEMA 2017).

The third objective is to **practice good governance**. The systems approach identified earlier mentions breaking down silos. Agencies and businesses can recognize their interdependencies with each other and collaborate on resilience building strategies. This involves building alliances across departments; coordinating communication channels and business functions; consolidating resources, etc. At the government level, leadership and their managers can work to formalize standards, processes, and procedures that guide operations and ultimately ensure continuity of operations. This also means that governments work to regulate and enforce policies that manage natural resources and development activities consistently and fairly.

As ambitious as these objectives might seem, most island communities would embrace them. Furthermore, so would government workers. The UVI team surveyed USVI government employees as part of the information gathering for the HMRP and learned that addressing these issues to make the government function more effectively could readily improve morale, performance, and retention among the public sector employees (HMRP, unpublished data).

In summary, building capacity to enhance climate change resilience in Caribbean island nations and territories is a difficult and complex challenge. In the USVI, by framing resilience in the context of hazard mitigation strategies and investment, decision-makers and governments can focus on much needed improvements in governance and regulatory policy implementation. Such efforts can be achieved by managing environmental monitoring information, fostering collaborations within and among agencies and their private sector partners, effectively communicating risks to communities, and sustainably managing natural resources. Collectively, the improvements could go a long way toward creating a work environment that encourages and recruits the technical expertise needed to meet the challenges of climate change in the Caribbean.

Acknowledgements We thank the Centre national de la recherche scientifique based in Paris, France, the US Army Corps of Engineers, and the National Oceanic and Atmospheric Administration of the USA for funding to support the workshop in Barbados in June 2022 which was the basis for the development of this book. We especially thank Ms. Shelly-Ann Cox for her help in organizing and running the workshop.

References

Banerjee O, Boyle K, Rogers CT, Cumberbatch J, Kanninen B, Lemay M, Schling M (2018) Estimating benefits of investing in resilience of coastal infrastructure in small island developing states: an application to Barbados. Mar Policy 90:78–87

Cambers G, Claro R, Juman R, Scott S (2008) Climate change impacts on coastal and marine biodiversity in the insular Caribbean. Caribbean Natural Resources Institute (CANARI), Port of Spain, Trinidad, Technical Report 382, 75 pp.

CCCCC (2012) Delivering transformational change 2011–2021: ìmplementing the CARICOM regional framework for achieving development resilient to climate change. Caribbean Community Climate Change Centre, Belmopan, Belize, 27 p. ///C:/Users/ozner/Downloads/Delivering%20Transformational%20Change%202011-21%20(Executive%20Summary)_%20Implementing%20the%20CARICOM%20'Regional%20F.pdf

Diakakis M, Deligiannakis G, Katsetsiadou K, Lekkas E (2015) Hurricane Sandy mortality in the Caribbean and continental North America. Disaster Prev Manag 24:132–140

ECLAC (2019) Report of the learning conference on implementing the sustainable development agenda in the Caribbean Region, LC/CAR/2019/3. Economic Commission for Latin America and the Caribbean, Santiago, Chile

FEMA (2017) Natural hazard Mitigation Saves. Federal Emergency Management Agency, Washington, D.C., USA. Interim report. https://www.wbdg.org/files/pdfs/MS2_2017Interim%20Report.pdf

Mercer J, Kelman I, Alfthan B, Kurvits T (2012) Ecosystem-based adaptation to climate change in Caribbean small island developing states: Integrating local and external knowledge. Sustainability 4:1908–1932

Muñoz S, Ötker I (2018) Building resilience to natural disasters in the Caribbean requires greater preparedness. International Monetary Fund, Washington, D.C., USA. https://www.imf.org/en/News/Articles/2018/12/07/NA120718-Building-Resilience-to-Natural-Disasters-in-Caribbean-Requires-Greater-Preparedness

Mycoo M (2014) Sustainable tourism, climate change and sea level rise adaptation policies in Barbados. Nat Res Forum 38:47–57

Resilientvi.org (2022) A hazard mitigation and resilience plan for the USVI. St. Thomas, US Virgin Islands. https://resilientvi.org/

Reyer CPO et al (2017) Climate change impacts in Latin America and the Caribbean and their implications for development. Reg Environ Change 17:1601–1621

Rhiney K (2015) Geographies of Caribbean vulnerability in a changing climate: issues and trends. Geogr Compass 9:97–114

Robinson S (2018) Adapting to climate change at the national level in Caribbean small island developing states. Island Stud J 13:79–100

Thomas A, Benjamin L (2018) Policies and mechanisms to address climate-induced migration and displacement in Pacific and Caribbean small island developing states. Int J Clim Change Strateg Manage 10:86–104

UN-OHRLLS (2011) Small island developing states: Small island, big(ger) stakes. United Nations Office of the High Representative for the Least Developed Countries, Landlocked Developing Countries and Small Island Developing States. New York, NY, USA. Available at: https://www.un.org/ohrlls/

Uyarra M, Côté I, Gill J, Tinch R, Viner D, Watkinson A (2005) Island-specific preferences of tourists for environmental features: implications of climate change for tourism-dependent states. Environ Conserv 32:11–19

Weaver DB (2001) Mass and alternative tourism in the Caribbean. In: Harrison D (ed) Tourism and the less developed world: issues and case studies. CABI Publishing, Oxford, U.K., pp 161–174

World Bank (2016) New country classifications by income level: 2016–2017. This online site is regularly updated on July 1 of each year to list per capita national income for countries. Available at: https://blogs.worldbank.org/opendata/new-country-classifications-2016

World Commission on Environment and Development (1987) Our common future. Oxford University Press, Oxford, U.K., 393 pp

Chapter 2
Principles of Systemic Resilience to Climate Change in Caribbean Small Island Developing States

Rebecca A. Rebar, Jose M. Palma-Oliveira, Gianluca Pescaroli, Gregory A. Kiker, Roxann K. Stennett-Brown, Chante Saunders, Gregory E. Guannel, Benjamin D. Trump, Igor Linkov, and James H. Lambert

Abstract It is widely accepted that climate change is associated with vulnerabilities to infrastructure, and to economic and societal risk across the globe. Developing island nations are more susceptible to climate change impacts due to the unique location and logistical challenges already faced by these countries that are

R. A. Rebar
Centre for Risk Management of Engineering Systems, University of Virginia, Charlottesville, VA, USA
e-mail: pcv9md@virginia.edu

J. M. Palma-Oliveira
Faculty of Psychology, CIPSI, University of Lisbon, Lisbon, Portugal
e-mail: joseo@campus.ulisboa.pt

G. Pescaroli
Institute for Risk and Disaster Reduction, University College London, London, UK
e-mail: g.pescaroli@ucl.ac.uk

G. A. Kiker
Department of Agricultural and Biological Engineering, University of Florida, Gainesville, FL, USA
e-mail: gkiker@ufl.edu

R. K. Stennett-Brown
Department of Engineering, University of West Indies, Mona Campus, Kingston, Jamaica
e-mail: roxann.stennett02@uwimona.edu.jm

C. Saunders
Center for Resource Management and Environmental Studies, University of West Indies, Cave Hill Campus, St. Michael, Barbados

G. E. Guannel
Caribbean Green Technology Centre, University of the Virgin Islands, St. Thomas, Virgin Islands (U.S.)

B. D. Trump · I. Linkov
Engineer Research and Development Center, US Army Corps of Engineers, Vicksburg, MS, USA

C. D. Metcalfe and E. R. Bennett (eds.), *Building Resilience to Climate Change in Small Island Developing States (SIDS) in the Caribbean*, Environmental Contamination Remediation and Management, https://doi.org/10.1007/978-3-031-37376-3_2

due to their geographical isolation and need for closed systemic functions. Actions to mitigate climate change require practical, holistic, strategized solutions to enhance the resilience of island nations. This chapter describes a framework of principles for Small Island Developing States (SIDS) in the Caribbean to incorporate and enhance infrastructure and societal resilience to address anticipated climate shifts. The framework can be adapted for other climate-affected regions that have similar characteristics to those of the Caribbean SIDS.

2.1 Introduction

Developing island nations, especially those in the Caribbean, are likely to see even greater impacts to sustainability, economics, and society due to climate change (Gallina et al. 2016; Thomas et al. 2018; Bonato et al. 2022). In 2017, hurricanes Irma and Maria devastated the island of Puerto Rico. The physical damage of these storms was compounded by an economic downturn and unreliable infrastructure (Bond et al. 2020). Five years later, global temperatures are predicted to rise another 2 °C, increasing sea water levels and affecting salinity measurements, and more frequent and intense storms are forecasted. As the climate continues to shift, Caribbean islands must develop effective plans for sustained economic growth and resilience (NOAA 2022, Mycoo 2018). Their preparation for continual climate shocks is vital for sustainable, long-term adaptations to various climate-related events anticipated to overwhelm these vulnerable habitats.

Despite this, adaptations to climate change require strategic prioritization to identify and implement effective solutions (Loose et al. 2022; Bonato et al. 2022). To meet these strategic goals, *resilience* must be considered as part of the system design for infrastructure, economy, and society at large. In this chapter, a framework of principles for small island nations to consider in the development of their infrastructure, policies, and economic plans is discussed. The ideas presented herein can enhance the resilience of island communities in preparation for, rather than a response to, climate change impacts. This chapter focuses on incorporating principles of resilience in the policies, practices, and management of social-environmental systems for Caribbean SIDS to consider for building a capacity for resilience.

J. H. Lambert (✉)
Centre for Risk Management of Engineering Systems, University of Virginia, Charlottesville, VA, USA
e-mail: lambert@virginia.edu

2.2 Background

As policy initiatives and strategic outcomes are developed, any SIDS must consider resilience from climate shifts. The United Nation's Sustainable Development Goals (SDGs) originally promulgated in 2015 state that:

> Climate change is one of the greatest challenges of our time and its adverse impacts undermine the ability of all countries to achieve sustainable development. Increases in global temperature, sea level rise, ocean acidification and other climate change impacts are seriously affecting coastal areas and low-lying coastal countries, including many least developed countries and small island developing States (p. 5).

Several Caribbean islands have made significant progress toward meeting the 2030 SDG objectives. For example, the Barbados Department of Emergency Management (2020) developed a Strategic Plan to prioritize disaster response efforts and management principles with specific goals identified as driving forces in their efforts. However, the document doesn't discuss actions to mitigate and prepare for the effects of climate change in the region. It simply outlines the proposed responses to those effects, omitting the adoption of resilience practices to maintain the island's economic and societal growth through the impending climate changes.

Numerous sources outline various definitions of "resilience" as it applies to various impacted outcomes. A widely accepted definition in terms of disaster resilience, provided by the National Academy of Sciences and further applied by Linkov and Palma-Oliveira (2017), is "the ability to plan and prepare for, absorb, recover from, and adapt to adverse events". Another description of resilience is the ability of a system to transform and "the speed of recovery" (Logan et al. 2022). Therefore, to be considered resilient, an entity must consider its ability to prepare for, adapt to, and rebound from catastrophic events. In terms of climate change resilience for SIDS, this presents an opportunity for incorporating risk management and resilience factors in the development of their infrastructure and land-use planning, logistics networks, technology integration, and governing processes. Linkov and Palma-Oliveira (2017) describe that "resilience strategies have the potential to radically change how a nation prepares itself for the potential disruptions of key services such as its energy, water, transportation, healthcare, communication, and financial services. When nations prepare for recovery from external shocks of a significant magnitude, resilience strategies must be considered".

The totality of this approach, integrating resilience concepts in the design of economic and cultural operating strategies, could be considered the *science of resilience* (Wu et al. 2022a). This science would be the application of a systemic approach used to understand the impacts of climate change effects and long-term adaptations in decision-making across policy development, infrastructure design, emergency management planning, project prioritization, and societal attitudes. Using this approach, resilience is integrated into all levels of society, thus enhancing a culture of preparedness and resilience.

This chapter thus discusses methods of applying the science of resilience in developing island nations. It introduces methodologies and considerations for SIDS to utilize or build upon as a framework for their development of infrastructure and policies. Various metrics discussed in this chapter can also be used to assess and measure resilience preparedness and resource needs. The focus of these applications is on the concerns specific to Caribbean SIDS. The framework is unique as it addresses not only the economic impacts on an island state but considers social contexts as an integral notion of resilience.

The organization of the chapter is as follows:

- Section 2.1 provides a framework to explore the complexities of economic, social, financial, and political systems and the relationships necessary to balance variables within the system to maintain stability due to low-level or high-level systemic disruptions.
- Section 2.2 will discuss metrics and the advantages and disadvantages of those currently in place to measure resilience.
- Section 2.3 outlines the importance of non-technological solutions, including nature-based and social applications, to build long-term sustainability and resilience of an island nation.
- Section 2.4 discusses early warning system applications and the effectiveness of these countermeasures.
- Section 2.5 describes the shortfalls of technology reliance specific to island nations and introduces methodologies to retain business continuity despite disruption.
- Section 2.6 describes the opportunities for ingenuity and innovation from SIDS resilience development and success as a model for global application.

2.3 Methods

The technical and scientific communities have built theories and mental models of complex systems in a way that elucidates the structure and processes to be analyzed and explained. The introduction of the concept of the Socio-ecological system (SES), and associated concepts like *resilience*, *panarchy*, or even the *adaptative cycle*, aims to examine these principles.

By definition, a *system*, given its adaptative cycle typically, is confronted with stressors that relate to a release or destructive phase. Recovering from the destructive phase can be faster and least prone to evoke human suffering and economic disruption, or in contrast, even evoke a profound *regime shift* where the system, as was before, can disappear and become a different one (Allen et al. 2014). The dynamic is complicated with the recognition that a system, after a stress or destructive phase, goes through reorganization and growth moments that, despite potentially bringing some novelty and diversity to the system, tend to be absorbed and regain the same configuration as before, including the same pitfalls (Pescaroli and Alexander 2016; Palma-Oliveira and Trump 2018). That is why even after disruptive events of a larger magnitude, the original systems do not necessarily change their function.

In the case of the Caribbean nations, the islands are faced with climate events that each must recover from, through *adaptation* of the system to a new dynamic or reabsorption of the impact into the existing structure. Thus, applying this adaptive framework to developing island states uncovers *systemic interactions* usually discovered by observing more common-sense hierarchal views that assume top-down decision-making.

The dynamics of various system entities, interactions between levels of the system's components, causes, cascading effects, and hierarchies of prioritization all affect system stability. For example, sea level rise mainly affects the coastline of an island, with a seemingly small overall impact. But coastal degradation shifts land use near coastlines, plausibly causing infrastructure damage, and increasing the likelihood of more severe weather effects in the area. The combination of these concerns thus alters economic and socio-economic decision-making for affected coastal areas. Island nations that rely heavily on the tourism industry for economic growth should consider shoreline erosion as a critical concern as it will impact shore-side structures such as oceanfront resorts, thus abating economic revenue. Considering sea level rise as an additional climate impact on shoreline degradation, the importance of accounting for both short-term and long-term resilience strategies becomes more crucial. Resource allocation and policy development must scrutinize all climate-related risks and balance immediate emergency response capabilities with longer-term climate adaptation needs.

It can be assumed that each island state is a *socio-ecological system* (SES), as defined by Ostrom (2007). The functions of each SES relate to another same-level SES (for example, other islands and/or a mainland region with similar socio-economic dynamics, such as Belize or Guyana) that provides different flows of people and goods, ultimately changing the SES of the island. The SES of each level is also connected with one or more higher-level SES that can make a long-lasting impact on an individual island state. An example of this interaction can be described by the first colonial occupation of Barbados that settled on the central areas and east coast of the island because of the specific psychological drives (i.e., agriculture). When those psychological drives changed the demand, a tourism-driven economy shifted to the west coast, which imparted a different array of implications for the SES of this island nation. The shift modified the way Barbados responds to natural hazards, making resilience more costly and difficult.

Also, within an island state, there exist multiple low levels of SES that have different echelons of resources and users and employ different types of governance and regulations. An example of this is that even small islands are distinguished by zones that differ in the characteristics of the coast and human occupation and the regulatory requirements required to manage each (CZMU 2020). A conclusion derived from comprehending these interconnected SES dynamics is that almost all SES have a set of factors that are permanent and those that can be described fully at a particular point in time. When the permanent state is understood, an almost complete description of the particular SES dynamics becomes apparent. A heuristic model for describing the SES (in the level of analysis necessary to meet the objective) was proposed by Ostrom (2007). From this SES interplay, it is concluded that resilience, as an emergent property of the system, can be evaluated when the system

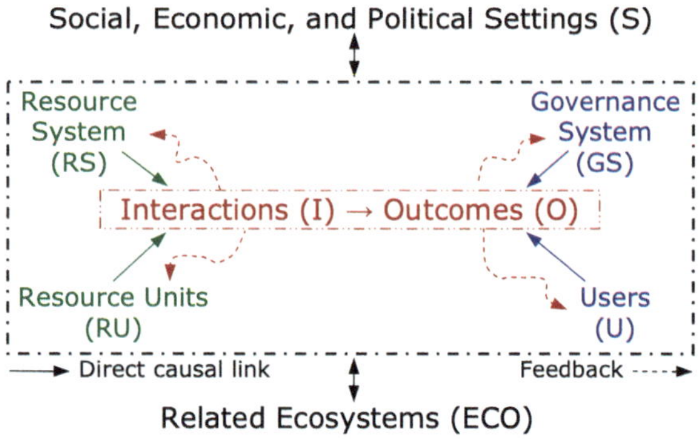

Fig. 2.1 A description of a Socio-Ecological System (SES) proposed by Ostrom (2007)

is confronted with a stressful event. The resulting outcome for the system varies due to the relationship among these variables (Fig. 2.1).

For example, a coastal zone can be profoundly changed in its capacity to absorb a stress due to several reasons. It can be predicted that the impact of a similar natural disaster would be higher on the modified, or urbanized, coastline (System A) than on the unmodified, natural coastline (System B). The *resilience* of the various SES is measured by the capacity of returning to the normal operation of the system. With some disasters, sometimes both types of systems can return to the same state as before, i.e., both are resilient. However, there still exist some crucial differences related to the type and resources required to build resilience. The resilience of the modified System A is maintained using the resources of a larger SES, i.e., international help and additional investments to retain continuity, typically a resilience strategy that involves greater or longer human suffering following an event. Furthermore, a specialized system typically operates mainly to meet the need(s) of a higher-level SES. For example, ecotourism is highly sensitive to disruptions in the larger, higher-level SES, such as a pandemic, or significant economic downturn.

Another implication of this systemic view of resilience is that the interplay between low-level and higher-level SES must be seen from a new perspective. Normally standard societal functions and their relationship with ecological systems are viewed as a hierarchical system. Control in a system is typically viewed as exerted by the larger entity in a "top-down" manner (Allen et al. 2014). Though this perception must be viewed in the opposite direction; that the control can be exerted by small-scale, "bottom-up" processes. Applying the concept of "panarchy" to this power dynamic can elucidate the practical and theoretical consequences of this view.

A *panarchy* is a nested set of socio-ecological systems in their different evolution points (considered as adaptative cycles) that are operating at different scales (Allen et al. 2014). Sometimes it is a larger-scale system that imposes the reorganization of an SES, as described in the resilience scenario of System A, necessitating higher-level intervention for successful recovery. For other SES, the modification of a small component within the system, for example, the destruction of coral reef barriers, promotes a regime shift in the structure or function of the system; thus, modifying the larger, higher-level SES (Cooper et al. 2020). This is especially impactful in the

SES of an island nation and important to note since the timescale for the bottom-up influences on system dynamics typically occurs at a rapid pace.

The implementation of the previously described set of principles is necessary to promote the long-term resilience of island nations. SES resilience in general must be viewed from a systems perspective where the critical function of an SES defines its limitations, as described by Linkov et al. (2014). For an island state, this is simplified when the natural SES is assumed to be the island. Thus, the definition of the island's SES hierarchy structure is straightforward and the connection with larger SES is easily identifiable. Using this framework and the identification of the system variables described above, the needs and drives of an island state can more easily be addressed.

In the case of Barbados, governmental and regulatory bodies are highly proficient at defining zoning procedures and internal management practices (CZMU 2020). However, since zoning procedures are tied to physical variables specific to the requisites of a particular region, the global dynamics and interactions within and between different SES levels fail to meet the island's needs. This contributes to why coordination failures abound where stakeholders have different needs and perspectives and are trying to push or contravene different actions. To combat this issue, a larger view of resilience across the entire system is necessary.

The complexities of this system outlook can be overcome with this heuristic perspective that allows for the possibility of stress tests. This perspective can make much clearer the understanding of the timing of a system(s) shock, the prioritization of needs, the presence and process of hierarchal structures with their real value and implications, and the overall coordination of resources. Temporal system shocks will have various levels of impact on the overall system balance, depending on the status of other variables in the system. For example, if a hurricane devastates an island's coastal zone while the island is experiencing extreme drought conditions, the priority for response will likely be different than under conditions where there are sufficient water resources. Thus, a strong overview of the vulnerabilities at the time of an event must be considered to ensure that the proper resources are provided.

Management processes and policies are critical to properly balancing the system. Regulatory agencies must consider response management concepts for both emergencies and long-term projects in terms of funding, resource allocation, and response procedures for the global SES and sub-systems (Palma-Oliveira and Linkov 2017). The system cannot return to an equilibrium that effectively prepares for long-term resilience, which is the goal, if it cannot respond efficiently to a short-term system shock. This can be seen in many variables but is more important in the social and economic sectors.

Building resilience is also not a responsibility solely held by government officials or industrial specialists. Citizens must also build internal resilience and adaptability to withstand regular shocks and the shifting of their livelihoods, especially due to the impacts of climate change. Trust in the decisions made by officials and industry must be established with the local community to avoid a "tragedy of the anticommons" as described by Palma-Oliveira et al. (2018). To avoid the anticommons tragedy requires that decisions are made in an effort to not overly burden one population, constituency, component, etc. of a system despite having a goal that benefits the entirety of the construct (Palma-Oliveira et al. 2018). In terms of climate change resilience, this means that decisions made to build resilient infrastructure, adaptive programs, and economic fortitude should not come at the expense of social and cultural degradation. Considerations for the community, using community-driven and led processes, are critical to successfully establish a resilient nation.

Finally, *hierarchical* but flexible structures must be developed so that the system can find equilibrium even in dire circumstances; that is, it should not be so highly regulated or "top-heavy" that actions cannot be taken without a single linchpin or failure halting future progress. Power and influence should be evenly distributed within and across the SES, allowing for multiple sectors to elevate or level others, depending on the inputs and events affecting the entirety of the system. Ensuring a hierarchical balance across the variables mitigates the potential for the issues with the *anticommons*, as previously described since it provides a calibrated approach across all sectors affecting the system's continuity.

The subsequent discussions relate to variables within the hierarchal structure of SES that, if enhanced, build resilience at critical junctions. These methods focus on the application of management principles that prepare for system-impact events and their interactions with social, environmental, and organizational components. The type of interconnectedness among these system variables defines the output of the system and its ability to retain operability, and thus be resilient.

2.4 Expand Disaster Metrics for Resilience

A key to understanding SES-affected variables is the ability to measure impacts on the system. Doing so requires a dynamic set of metrics that are scalable, broadly applicable, and can provide both an instantaneous value and a predictive or potential value. Specific to SIDS, metrics used to evaluate disaster risk, impact, and resilience are critical to receiving adequate resources for long-term success and achieving balance in the previously described system. In this section, current climate-related metrics for disaster valuation are discussed as they relate to the unique challenges that island nations face.

The scale of impact of disasters related to climate change in island nations is far greater than in larger geographic regions that typically have more varied economic portfolios. In 2020, Hurricane Dorian, a category 5 hurricane, caused USD $3.4 billion of damage in the Bahamas in less than 48 h (IDB 2020). Another storm of category 5 intensity, Hurricane Michael, hit the Florida panhandle in 2018, causing USD $25 billion of damage as it swept across the southeastern United States (Beven et al. 2019). Despite both being classified as category 5 hurricanes, the damage to the U.S. coastline costs more than 6 times that of the Bahamas. Yet the damage in the Bahamas equated to a loss of 25% annual GDP, while the impact on the annual GDP of the USA was less than 1% (IDB 2020). The only relatable example of a single event causing this GDP decline in the USA was the second quarter of 2020 when the COVID-19 pandemic costs 32.9% of the annual GDP over the course of three months

(Horsley 2020). As seen in this example, the commonly used metrics of GDP and current dollar value to measure disaster impacts is not adequate to understand and relate the totality of climate-related events, and more broad, dynamic, and system component-related measures must be identified.

An accurate evaluation of these climate-related impacts is necessary to ensure proper resource distribution and preparation for climate change events from a global perspective. Appropriate comparisons of impacts across nations must be evenly measured to determine the status of a system's balance across the resilience spectrum. The right metrics also determine the long-term impacts of short-term shocks impact over time.

Considering that the scope and complexity of GDP may not be an appropriate holistic measure of disaster impacts, other broadly applicable metrics must instead be identified. Yew et al. (2019) identified 15 vulnerability metrics and two exposure metrics to compare disaster impacts through a Disaster Severity Index (DSI). Various aspects of disaster response were selected, for the recoverability to be determined shortly after a system shock occurs. The metrics identified by Yew et al. (2019) listed in Fig. 2.2 provide a holistic picture of the total impact to an area. In comparison to the GDP metric, the DSI provides a more detailed, area-specific disaster comparison that allows for scalability.

The DSI identifies categories of the most important vulnerabilities to address in a system immediately following an event to ensure quick response and ideally recovery. However, the current metrics do not identify the long-term impacts of the disaster, nor are they designed for pre-disaster resilience assessments. But, expanding some of the identified DSI metrics can expose gaps in preparedness to reinforce local, regional, or national level responses. Application of the expanded metrics pre, or post-event, could be utilized as a semi-quantifiable resilience assessment tool that identifies system risks to be addressed.

Expanding or adding principles to these original metrics can be used to consider the parameters for long-term resilience. One of the ways to adjust this table to better quantify the long-term resilience of island nations is to expand category 7, "Main source of economy impact site". Another metric to understand the full impact of an event is necessary, as many islands have low economic outputs due to limited resource availability. The "Distance to Radius" category should include an option for total devastation or destruction, as could be the case in Caribbean islands due to the impacts of hurricanes or tropical storms (e.g., The impact of Hurricane Dorian on several islands in the Bahamas in 2019 and the impact of Hurricane Maria on Dominica in 2016). Regarding the population-related categories, a total population count should be added to understand the proportion of the population affected when determining social impacts in categories 6, 16, and 17.

Several other factors could be added to the table to understand the resilience of an island at any time. A key factor that is not currently considered is the availability of electricity as it is a key component of communications, healthcare, and supply chain continuity. This is especially important post-event but can also be a cause for concern for regular business continuity. Another consideration should be in place for noting if the event is part of a "compound casualty". For instance, if a major

Fig. 2.2 Vulnerability Indicators Criteria developed by Yew et al. (2019). These metrics can be expanded and used to measure long-term resilience in a system

15 Vulnerability indicators	**Criteria**	**Score**
1.Time Occurrence	Morning (06:00-14:00)	1
	Afternoon or late evening (14:00-22:00)	3
	Night (22:00-06:00)	5
2.Impact Time/Duration	1-24 hours	1
	>24-72 hours	3
	>72hours	5
3.Topography	Planar	1
	Mountainous	3
	Isolated island	5
4.Radius from the impact site	1-10km (1district/town/village)	1
	>10km (>1 district/town)	3
	>100km (>10 districts/towns approximately)	5
5.Accessibility to the impact site	No disruption in main transport lifeline (road, air, sea)	0
	Accessible via land(motorbike/bicycle)	1
	Accessible via land(animals/by walking)	2
	Accessible via sea route (boat) & air (helicopter)	3
	Accessible via air (helicopter)	4
	Not accessible at all to the impact site	5
6.Population Density	Low (rural)	1
	Medium	3
	High (urban)	5
7.Main source of economy at impact site	not affected	1
	temporary affected	3
	long term affected	5
8.Public infrastructure	Public infrastructure not affected (no changes before/after disaster)	1
	Public infrastructure <25% affected but still functioning	2
	Public infrastructure <50% affected but still functioning	3
	Public infrastructure >50% affected, need assistance	4
	Public infrastructure totally affected , total assistance needed	5
9.Communication	Communication not affected (no changes before/after disaster)	1
	Communication partially affected, only able to communicate via radio satellite phone	3
	Total disruption of communication network	5
10.Type of country economic	high income country	1
	middle income country	3
	low income country	5
11.Governance	**Corruption Perception Index 2015 Ranking (168 countries),to be adapted as per latest ranking**	
*in the case of countries ranking not		
available, global territories	0-20	5
average ranking will be selected.	>20-40	4
0= highly corrupt	>40-60	3
100=very clean	>60-80	2
<50=serious corruption problem	>80-100	1
	Basic Survival Resources: Water and Sanitation, Food Security, Shelter	
12.Water	No disruption in water infrastructure	0
	Disruption in water infrastructure	1
	Available of water source	2
and	Access to clean water via purification tablets/ceramic filter	3
	Access to clean water via drilling bore hole well	4
	Non-potable water/Not available water source and increased in diarrhea,skin & water related diseases	5
Sanitation Hygiene	No disruption in sanitation infrastructure	0
*Water & Sanitation scoring= (total water +sanitation hygiene	Disruption in sanitation infrastructure	1
scoring)	Available of mobile/ portable latrines ratio 1:<20	2
/2 to get the average	Available of mobile/ portable latrines ratio 1:20	3
	Available of some sort of damaged sanitation	4
	Open defecation & increased in diarrhoea & communicable diseases	5
13.Food Security	Food available & accessible (No changes before /after the disaster)	1
	Food available but needs transport from outside the impact site	2
	Food available but not accessible due to logistical & security issues	3
	Food stocks severely depleted (food scarcity)	4
	Food not available in or to the impact site.	5
14.Shelter	Shelter intact	1
	Shelter <25% destroyed	2
	<50% of the total shelter destroyed	3
	>50% of the total shelter destroyed	4
	All shelter totally destroyed	5
15.Healthcare Capacity	able to cope - all health facilities intact & functioning as normal situation	1
	- <25% of the total health facilities destroyed	2
	- <50% of the total health facilities destroyed	3
	unable to cope - >50% of the total health facilities destroyed	4
	- All health facilities totally destroyed	5
	Total 15 vulnerability indicators score	
2 Exposure indicators	**Criteria**	**Score**
16.Number of Deaths	10-100	0
	>100-1000	1
*at least 10 or more= 0	>1000-10 000	2
therefore less than 10 = 0 score	>10 000-100 000	3
	>100 000-1 000 000	4
	>1 000 000	5
17.Number of Affected Persons	100-1000	0
	>1000- 10 000	1
	>10 000- 100 000	2
*at least 100 or more= 0	>100 000- 1 000 000	3
therefore <100=0	>1 000 000- 10 000 000	4
	>100 000 000	5
	Total 2 exposure indicators score	

hurricane makes landfall during a pandemic, this can significantly affect the ability of an already over-burdened sector of the system to recover (Pescaroli and Alexander 2016). This consideration could also include a temporal component that considers the time elapsed since the last event to measure societal familiarity with the response process. This would also assess the response readiness from a societal aspect. Factors in this category would include relevant and recent response training, familiarity with disaster management policies and readiness scenarios, and prior preparation for an event, such as early warning procedures and early resource distribution. Including these factors in the overall scale better identifies the state of readiness within the affected system pre-event, providing a baseline to compare overall impacts to the system at a quick glance, and allowing for easier identification of the most pressing issues for building overall resilience.

2.5 Assets for Building Resilience

Resilience capabilities are often regarded in the preparation for a disruptive event, typically associated with the quantification of resource availability, the capacity of a system to withstand or recover from an event quickly, or the ability to withstand diverse threats (Galaitsi et al. 2021). While these themes resonate within resilience concepts for single-event disruptions, resilience on a societal level must include long-term preparations that are not directly linked to a specific disaster, but broadly appropriate measures for disaster preparedness. Considering this concept, the capacity of natural defenses to enhance resilience capabilities must be considered. In this context, non-technical solutions and the skills of a population become vital assets for climate change response, preparation, and long-term resilience.

Over-reliance on technology is a common issue in developing resilience response practices. Though typically advantageous at enhancing societal advancement when properly functioning, technological solutions do not have a zero percent chance of failure. Oftentimes, a failure of technology causes more disruptions to system operations than if it wasn't included as a critical component. Thus, technological solutions cannot guarantee reliability for climate change initiatives and alternative solutions that incorporate non-technological means should be considered.

A top, non-technical approach for preparing for climate change is the inclusion of nature-based solutions (NbS) for building resilience. Seddon (2022) describes NbS as "actions that involve people working with nature, as part of nature, to address societal challenges, providing benefits for both human well-being and biodiversity". Integration of NbS as a climate change mitigation strategy has many socio-economic impacts including ecology diversification, climate change impact mitigation, and societal involvement and maintenance for long-term implementation in an SES (Turner et al. 2022). Thus, including NbS builds multiple levels of resilience as these solutions include variables that enhance not just physical resilience, but also social, economic, and environmental resilience as well. Other chapters in this book describe case histories related to the application of NbS in Caribbean SIDS.

In terms of climate change resilience, NbS provide alternatives to engineered or "gray" shoreline infrastructure designed to protect or shield coastal environments. Guannel et al. (2016) describe how the interaction of mangroves, coral reefs, and seagrasses provides a natural defense to coastal erosion under baseline conditions, but also enhances shoreline stability during storm conditions. Coastline assessments also show that natural habitat conservation and restoration of natural habitats reduce by half the exposure to hazards in coastal communities (Arkema et al. 2013).

The importance of utilizing non-technological solutions as a resilience strategy for combatting the impacts of climate change is growing as more studies contribute to understanding the role of natural defense strategies. One such study seeks to understand the inundation of tsunami and storm waves using subtidal coastal boulder deposits as indicators of past storm impacts and shifts for the future (Kennedy et al. 2021). Natural defense mechanisms can be incorporated to supplement technological solutions (i.e., "hybrid" solutions) in preparing for and withstanding the long-term effects of climate change. Sustaining NbS ensures that the success of shoreline protection systems is not hinging on a single point of failure within a technological solution. Rather, NbS allow for the diversification of a climate-resilient society and system.

Societal preparedness is an aspect of resilience that is often overlooked. MacLean et al. (2013) described social resilience with six key attributes that are also linked to the economy and environment which describe the balance for system resilience in the context of an SES. These factors include knowledge, skills and learning, community networks, people–place connections, community infrastructure, a diverse and innovative economy, and engaged governance (MacLean et al. 2013). SIDS are especially reliant on utilizing their people as a force for building long-term resilience, because alternative climate adaptations, for example diversifying economic output, are more difficult because of the geographic challenges of island economies.

Relating the concept of social resilience to the SES described in Fig. 2.1, these attributes can be incorporated into the system of resources, governance principles, and stakeholders. Examples of implementing social resilience can be seen in a variety of methods already employed. One application of this is education. The Virgin Islands Established Program to Stimulate Competitive Research (VI-EPSCoR) established climate-focused education opportunities for students to develop a foundational understanding of climate-related issues affecting the Virgin Islands and encourages community engagement in implementing and sharing climate resilience information, projects, and planning (VI-EPSCOR, n.d.). Additionally, student outreach and activities help to understand climate impacts more thoroughly (Wedding et al. 2018). Some of these capacity building approaches are described in Chap. 1 of this book.

Another application of the above principles is by preparing for emergency situations through policy development; conducting mock scenarios to understand the effectiveness of policy requirements and resource availability. Linkov et al. (2022) describe a similar approach as *stress testing* of critical infrastructure. Learning the limitations of system resilience requires engaged governance which enhances understanding and the connectedness of inter-related components, thus increasing social resilience factors.

Preparing society for climate-related adaptations and developing coordinated, effective responses ensures society can regain balance across the SES faster for long-term continuity and operability. It shifts the resilience focus to a long-term sustainability ethos. It also enhances resilience by expanding the impact of tools and resources already available within the system.

2.6 Early Warning System Applications

In addition to added resilience by utilizing physical, natural solutions, properly implemented technological systems can also enhance preparedness. Implementation of some technological approaches is critical to island communities that often face the potential for natural disasters. One such climate-related technological solution is the effectiveness of Early Warning Systems (EWS) in the Caribbean islands.

Weather-related early warning systems continue to expand as scientific research discovers earlier signs that warn of many impending natural phenomena. While many Americans are familiar with hurricane warnings via news and other sources, other EWS focus on a variety of natural disasters associated with a region. In the context of islands, several EWS are in place to identify the onset of earthquakes, landslides, tsunamis, coastal flooding, wildfire spread, and droughts (Argyroudis et al. 2022). Although these tools can be useful in emergencies, the reliability, familiarity, and trust of these systems must be developed to ensure that the tools are properly employed.

To ensure the successful management of disaster preparedness, a strong understanding of the risks of the event, including the impact of an action on the system, and an understanding of the vulnerabilities of the impacted region are critical (Hissel et al. 2014). The risks to Caribbean SIDS are similar across islands: Climate change impacts, exposure to severe weather, higher risks of isolation from external resources, and high potential for catastrophic impacts from one major weather-related event. Each island in the Caribbean has distinctive terrain, agricultural, and coastal profiles, as well as its own government policies, social practices, and economic strengths and weaknesses. Thus, each island is unique in the procedural deployment of EWS and its structure for implementation.

The relative importance of an EWS is described by Velazquez et al. (2020) in terms of four impact domains: operational, political and governance, social and behavioral, and organizational, which can be expanded from the context of earthquake-specific warnings to the broad category of EWS in general. In application, this relates to the parameters of the system, how the system is tested and employed, the public understanding and response to the system once enacted, and the utilization of the system in the disaster response. Figure 2.3 shows how several parameters promote system effectiveness.

Encouraging EWS effectiveness in each of these domains can be extensive. Involvement in the local community and an assessment of the attitude and persuasiveness of an EWS in the region must be known. Culturally, an attitude of resistance can

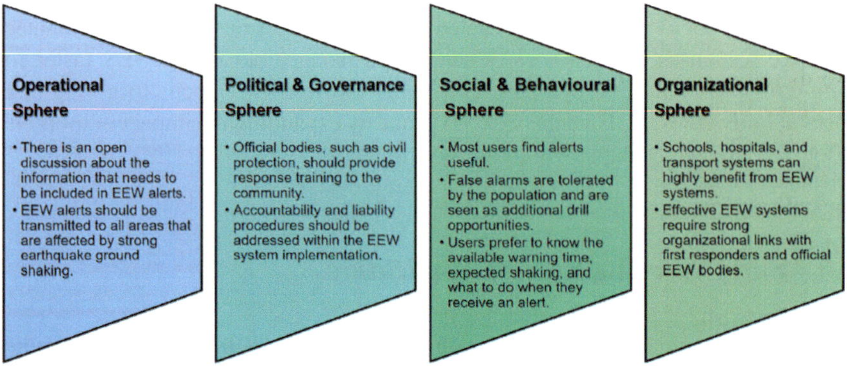

Fig. 2.3 Earthquake Early Warning (EEW) implementation challenges across several domains, which can be applied to multiple early warning system applications (adapted from Velazquez et al. 2020)

impede the effectiveness of EWS, regardless of the data that supports its need. Typically, this can be addressed through education and involvement of the community in the process of EWS execution. Awareness of the social perception toward EWS and other technological support systems can help to avoid anticommons concerns within communities.

Supporting this concept of system trust, an EWS must be accurate and effective at communicating the impending hazard. This requires allocating resources to implement new technologies, adopting new methodologies into preparedness research, and adapting the system to incorporate new threats. Finally, system deployment must be well-understood and exercised to ensure success. This relates to the concept of *stress testing* the system (Linkov et al. 2022). For example, the last hurricane to significantly affect Barbados struck in October of 2010, bringing heavy rain and damaging winds (NASA 2010). Earthquakes are a regular occurrence in some eastern Caribbean islands due to their proximity to an active tectonic plate boundary (Woods Hole 2018). Reports of the most recent tsunamis impacting the island of Hispaniola (i.e., Haiti and the Dominican Republic) were recorded in the mid-1900s (Woods Hole 2018). As urbanization, tourism, and economic growth have altered the socio-economic conditions of Caribbean SIDS since then, the need to regularly exercise these systems and prepare residents for the impacts, and to do so effectively, is essential to building social resilience.

2.7 Impacts of Technology Integration

Despite the traditional thinking that SIDS typically rely on low-technology initiatives and self-reliance to build resilience, the incorporation of these nations into the global economy has significantly enhanced their ability to implement organizational and technological solutions. Because of this new reality, it is important to highlight and assess the strengths and weaknesses of these interactions.

The incorporation of technology initiatives into infrastructure, operations, and standard business practices has reduced the isolation of SIDS from the global economy. This provides these island communities with the opportunities to develop their own sense of identity and integration into the international market. Although this integration affords SIDS the opportunity to expand their economies, providing competitive sources of supply and alternate sources of demand on a global scale, the inclusion of technologies into critical components of regular operations also increases the vulnerabilities by expanding external dependencies onto critical networks such as electrical grids, information and communication technology (ICT), transportation, and navigation capabilities. These dependencies are now essential for continued supply chain operations, government operations, and societal quality of life.

Over-reliance on certain technologies, though, can have devastating effects. For instance, if emergency management of a SIDS relies heavily on satellite or Internet communications for disaster response at a time when those capabilities are not available, the entire response system could collapse. In island nations, where logistics and response for emergency management must be self-reliant, underestimating the impacts of disruption or not identifying critical weaknesses could devastate a region or population. Mis-identification of critical points of failure within a system without preparation for system continuity and without the availability of technology or vital system components inhibits systemic resilience. This is especially true during times when the SES is already in a state of stress due to non-climate-related concerns such as during a pandemic or an economic recession.

Though critical infrastructure risks and resilience strategies have been studied internationally, the dependencies of SIDS on these integrated technologies have yet to be thoroughly examined. In terms of climate change impacts, the long-term effects of the extended losses of these key technologies are poorly understood, though indicators of potential outcomes are present from various examples of Caribbean SIDS. An example is the extended power outages and unstable power supply grid in Puerto Rico and its effect on the ability of the island to cope with extreme climate-related events. In 2017, Hurricanes Irma and Maria made landfall in Puerto Rico less than two weeks apart, compounding the challenges for emergency response, but causing catastrophic damage to the power grid. Impacts of island-wide power outage after these storms included more than 2,000 deaths not directly related to the causal event (Commonwealth of Puerto Rico 2018). Another example is the limited transportation capabilities in Haiti due to the frequency of natural disasters devastating the island (Ndambuki and Al Hitmi 2022). Lastly, a recent Internet outage in the South Pacific island of Tonga demonstrates the extent of societal impacts that can be triggered by the ICT failures of an island nation that heavily relies on international communications to conduct regular business (Al Jazeera 2019). Similar issues could result from malfunctions or reception of global navigation satellite systems, where an unseen utility affects all other services, such as the ability to conduct regular trade like maritime shipping or air travel.

Island nations also have the additional burden of understanding inter-related casualty factors. The connectedness of extreme weather events and potential technology disruptions makes Caribbean SIDS especially vulnerable to cascading casualties. Understanding the impacts of the incorporation of technology deeply within a vulnerable system can have grave consequences, especially if occurring in conjunction with or because of, another unrelated casualty. Pescaroli and Alexander (2018) describe this phenomenon as cascading casualties or "… the manifestation of vulnerabilities accumulated at different scales, including socio-technological drivers. The possible environmental triggers… can be associated with compounding and interconnected risk, while critical infrastructure and complex adaptive systems may be the drivers that amplify the impacts of the cascade". The assessment of possible gaps, identification of singular points of failure, and the impacts of either being exploited must be conducted during the development of contingency plans. As SIDS economies heavily rely on tourism and regular trade for economic prosperity, these two areas should be scrutinized for weaknesses and alternative measures for spanning potential gaps to enhance the system's overall resilience.

Routine contingency management practices fail to incorporate cross-border, multi-hierarchical scenarios. Takeda et al. (2017) described the necessity for local agents to be trained and able to make rational decisions regarding resource allocation, the needs of the community, and information gathering to effectively respond to emergencies with complex interactions and high uncertainty. Response methodologies and standard practices for overlapping technological and nature-based disasters are not readily available, which also limits understanding of how different types of casualties can affect one another, directly or indirectly. Building disaster response plans, which can be "stress-tested" with these "unusual" circumstances, can characterize the resilience and robustness of a region. Pescaroli et al. (2018) describe this scenario as the "*massive, overwhelming, disruption of operations* (Mo.OR.D.OR.)", during which accumulated vulnerabilities between natural and man-made risks can escalate a cascading disaster (2018). Preparing for this kind of contingency response requires significant planning, investment in resources, and a strong understanding of the interconnectedness of critical systems to build organizational resilience and continuity.

2.8 Innovation and Enterprise

SIDS in the Caribbean are in a unique position where the timeline of their governance development and infrastructure planning is relatively new. Because of this, these islands are primed for including resilience practices into their standard policies, procedures, planning, and management principles. The inclusion of resilience factors in the system's design supports enterprise continuity.

Prioritization of projects in decision-making is an example of including resilience in practice. Typically, the lowest-cost project is the priority. Through the use of multi-criteria decision analysis (MCDA), various projects can be compared to determine which solutions best meet the system goals and provide the most resilience in the system. This process has been demonstrated several times across various projects. Karvetski et al. (2011) assessed the impacts of sea level rise against various criteria goals, such as protection of public infrastructure versus protection against storm surge and floods. You et al. (2014a, b) applied the same concept to review transportation projects for prioritization of roadways in urbanized areas under various climate change scenarios. For developing countries, the ability to effectively prioritize projects is key to building critical infrastructure that aligns with the needs of the country. This process has been applied in multiple case studies where critical infrastructure is evaluated over emergent conditions such as infrastructure development and initiatives for building climate change resilience (Bonato et al. 2022; Lambert et al. 2013). Using MCDA principles in risk prioritization, especially related to a developing nation overcoming the uncertainty of climate change impacts, decision-makers can understand the trade-offs when choosing to pursue one project over another. MCDA can also assist with resource allocation and balancing stakeholder needs, including public opinion.

A key part of this process is knowing the system goals. Galaitsi et al. (2021) described the need to understand system performance concepts to "pursue management strategies better suited to their unique system dynamics and specific objectives of good performance". System functions change over time and the timing of an impact to the system can alter the system's operational performance, and also the trajectory of its long-term success. Having clear definitions of system success, such as resilience, adaptability, or sustainability, requires that decision-makers align their governance practices to meet those objectives (Galaitsi et al. 2021). The principles previously discussed in this chapter are applicable considerations, especially the need to "stress test the system", for examining alignment with the stated system goals.

Caribbean SIDS have an incentive to develop resilience within their island communities across all domains. Because these islands are highly influenced by the effects of climate change, they are acutely aware of the necessity of building resilience into the social, economic, and environmental sectors of their nations. Thus, the goals can be identified, project values established and prioritized, and decision analysis processes completed to determine the best direction to reach system success.

2.9 Discussion

In this chapter, we have described how socio-economic system variables, hierarchy, and interactions of SIDS can be addressed to build resilience to climate change. Application of the above principles allows SIDS to reflect on the nature of the current SES and identify opportunities to build resilience. Though post-event resilience and rebuilding will always be applicable, preparation for system shocks can decrease impacts on human suffering and mitigate the costs to rebuild (Palma-Oliveira and Linkov 2017). With this preparation, non-technological approaches and building social resilience must be a focus. SIDS must also evaluate the state of readiness of their communities and know the capabilities for disaster response and recovery to effectively allocate resources for long-term continuity, but also in emergency situations. Stress-testing the capabilities, building multi-level, cross-domain socio-ecological infrastructure, and leveraging technological and natural innovations can help to achieve maximum systemic resilience.

The whole of these methodologies culminates in a few overarching themes. Notably, each section introduced the concept of a component or system breakdown. This leads to the appreciation of how such failures might be inevitable. Planning for, and understanding the impact of, a particular failure and its total effect on the current state of the system is vital to building resilience within the whole enterprise. Another concept is the cyclical nature of resilience within systems. Resilience is dynamic, with multiple time horizons. An instantaneous measure of resilience must be monitored, re-evaluated, and re-calibrated as both external and internal influences shift priorities, and as a result, more or fewer tools are available to maintain the same level of operations for continuity. Finally, all the tools introduced in this chapter revolve around the principle of building a resilient society. This requires that communities can adapt to changing environments: political, ecological, and otherwise. It also requires that at a societal level there is a strong balance of understanding and trust in building the interconnectedness of critical systems necessary for resilience. Incorporation of these methodologies allows SIDS to.

- understand system dynamics and the several time horizons,
- accurately assess resilience capabilities,
- leverage non-technical solutions for building resilience,
- understand the advantages and limitations of technology in resilience capabilities, and
- utilize case histories of successes from Caribbean SIDS to build global enterprise resilience.

Solutions to multi-faceted problems are ultimately complex and cannot be completely addressed from a single perspective, single system review (i.e., social vs. economic vs. environmental), or a single point in time. The response must be adaptive, malleable to the system dynamics and shifts over time, and holistic in the approach and review.

2.10 Conclusions

In this chapter, we described principles and a framework for how building resilience in Caribbean SIDS is possible with a dynamic balancing of factors from several technology and social perspectives. Preparation for events that disrupt balance, including multi-level and short- and long-term impacts, is critical to developing a system that is able to recover from disruptions, thus achieving resilience. Pre-emptive actions require foresight of climate change impacts and anticipation of failures of critical technologies. Incorporation of NbS and the utilization and training with knowledge across disciplines are needed. Resilience to disruption is measurable across a range of scales and uses effective metrics for systemic resource allocation (Zhuang et al. 2022; Wu et al. 2022b). The continuity of operations through climate change is key to the sustained economic growth of SIDS and as a model for building resilience principles into the design of a nation.

Acknowledgements The authors would like to thank the US Army Corps of Engineers, the National Oceanic and Atmospheric Administration, and the Centre national de la recherche scientifique based in Paris, France, for sponsoring the "Resilience to Climate Change in Small Island Developing States in the Caribbean" Workshop. The authors are grateful in particular to Erin Bennett and Chris Metcalfe from Trent University for their organization of the event that allowed the Working Group to develop and discuss the material of this chapter. The authors are grateful for the support of the Commonwealth Center for Advanced Logistics Systems based in Petersburg, VA, USA.

References

Al Jazeera (2019) Tonga facing 'absolute disaster' after internet cable blackout. News. https://www.aljazeera.com/news/2019/1/23/tonga-facing-absolute-disaster-after-internet-cable-blackout

Allen CR, Angeler DG, Garmestani AS, Garmestony MS, Gunderson LH, Holling CS (2014) Panarchy: theory and application. Ecosystems 17:578–589. https://doi.org/10.1007/s10021-013-9744-2

Argyroudis SA, Mitoulis SA, Chatzi E, Baker JW, Brilakis I, Gkoumas K, Vousdoukas M, Hynes W, Carluccio S, Keou O, Frangopol DM, Linkov I (2022) Digital technologies can enhance climate resilience of critical infrastructure. Clim Risk Manag 35:100387. https://doi.org/10.1016/j.crm.2021.100387

Arkema K, Guannel G, Verutes G, Wood SA, Guerry A, Ruckelshaus M, Karieva P, Lacayo M, Silver JM (2013) Coastal habitats shield people and property from sea-level rise and storms. Nat Clim Chang 3:13–918. https://doi.org/10.1038/nclimate1944

Barbados Department of Emergency Management (2020) Strategic plan: 2019–2023. https://dem.gov.bb/public/downloads/DEM_%20Strategic_Plan_design_final_26March2020.pdf

Beven II JL, Berg R, Hagen A (2019) Hurricane Michael: National Hurricane Center Tropical Cyclone Report, National Hurricane Center, Miami, FL, USA. https://www.nhc.noaa.gov/data/tcr/AL142018_Michael.pdf

Biondi EL, Guannel G (2018) Practical tools for quantitative analysis of coastal vulnerability and sea level rise impacts—application in a Caribbean island and assessment of the 1.5 °C threshold. Reg Environ Change 18:2227–2236. https://doi.org/10.1007/s10113-018-1397-4

Bonato M, Sambo B, Sperotto A, Lambert JH, Linkov I, Critto A, Torresan S, Marcomini A (2022) Prioritization of resilience initiatives for climate-related disasters in the Metropolitan City of Venice. Risk Anal 42:931–952. https://doi.org/10.1111/risa.13823

Bond CA, Strong A, Smith TD, Andrew M, Crown JS, Edwards KA, Gonzalez GC, Gutierrez IA, Kendrick L, Louto JE (2020) Challenges and opportunities for the Puerto Rico economy. https://www.rand.org/pubs/research_reports/RR2600.html

CZMU (2020) Updated draft integrated coastal management plan. Coast Zone Manage Unit, Barbados. http://www.coastal.gov.bb/ICZMPlan

Commonwealth of Puerto Rico (2018) Puerto rico disaster recovery action plan. https://www.cleanegroup.org/wp-content/uploads/PR-Disaster-Recovery-Action-Plan.pdf

Cooper GS, Willcock S, Dearing JA (2020) Regime shifts occur disproportionately faster in larger ecosystems. Nat Commun 11:1175. https://doi.org/10.1038/s41467-020-15029-x

Galaitsi SE, Keisler JM, Trump BD, Linkov I (2021) The need to reconcile concepts that characterize systems facing threats. Risk Anal 41:3–16. https://doi.org/10.1111/risa.13577

Gallina V, Torresan S, Critto A, Sperotto A, Glade T, Marcomini A (2016) A review of multi-risk methodologies for natural hazards: consequences and challenges for a climate change impact assessment. J Environ Manage 168:123–132. https://doi.org/10.1016/j.jenvman.2015.11.011

Guannel G, Arkema K, Ruggiero P, Verutes G (2016) The power of three: coral reefs, seagrasses and mangroves protect coastal regions and increase their resilience. PLoS ONE 11:e0158094. https://doi.org/10.1371/journal.pone.0158094

Hissel F, Morel G, Pescaroli G, Graaff H, Felts D, Pietrantoni D (2014) Early warning and mass evacuation in coastal cities. Coast Eng 87:193–204. https://doi.org/10.1016/j.coastaleng.2013.11.015

Horsley S (2020) 3 Months Of Hell: U.S. Economy Drops 32.9% In Worst GDP Report Ever. https://www.npr.org/sections/coronavirus-live-updates/2020/07/30/896714437/3-months-of-hell-u-s-economys-worstquarter-ever

IDB (2020) Impact of Hurricane Dorian in the Bahamas: a view from the sky. Inter-American Development Bank, Technical Note N° IDB-TN-1857

Karvetski CW, Lambert JH, Keisler JM, Linkov I (2011) Integration of decision analysis and scenario planning for coastal engineering and climate change. IEEE Trans Syst, Man, Cybern Part A: Syst Hum 41:63–73. https://doi.org/10.1109/TSMCA.2010.2055154

Kennedy A, Cox R, Berke M, Guannel G, Lau A, Engel M, Mori N (2021). Coastal boulder deposits as a record of extreme inundation: the inundation signatures on rocky coastlines (ISROC) Research Coordination Network. AGU Fall Meet Abstr. https://ui.adsabs.harvard.edu/abs/2021AGUFMNH23A..07K

Lambert JH, Wu Y, You H, Clarens A, Smith B (2013) Climate change influence on priority setting for transportation infrastructure assets. J Infrastruct Syst 19:36–46. https://doi.org/10.1061/(ASCE)IS.1943-555X.0000094

Linkov I, Bridges T, Creutzig F, Decker J, Fox-Lent C, Kröger W, Lambert J, Levermann A, Montreuil B, Nathwani J, Nyer R, Renn O, Scharte B, Scheffler A, Schreurs M, Thiel-Clemen T (2014) Changing the resilience paradigm. Nat Clim Chang 4:407–409. https://doi.org/10.1038/nclimate2227

Linkov I, Trump B, Trump J, Pescaroli G, Mavrodieva A, Panda A (2022) Stress-test the resilience of critical infrastructure. Nature 603(7902):578

Logan TM, Aven T, Guikema SD, Flage R (2022) Risk science offers an integrated approach to resilience. Nat Sustain 5(9):741–748. https://www.nature.com/articles/s41893-022-00893-w

Loose DC, Eddy T, Polmateer TL, Manasco MC, Moghadasi N, Lambert JH (2022). Managing pandemic resilience with other cascading disruptions of a socio-technical system. In: Proceedings of the IEEE international systems conference (SysCon), pp 1–6

Maclean K, Cuthill M, Helen R (2013) Six attributes of social resilience. J Environ Plan Manage 57:144–156. https://doi.org/10.1080/09640568.2013.763774

Mycoo MA (2018) Beyond 1.5 °C: vulnerabilities and adaptation strategies for Caribbean Small Island developing states. Reg Environ Change 18:2341–2353. https://doi.org/10.1007/s10113-017-1248-8

NASA (2010) Hurricane Season 2010: Hurricane Tomas (Atlantic Ocean), National Aeronautics and Space Administration. https://www.nasa.gov/mission_pages/hurricanes/archives/2010/h2010_Tomas-Atlantic.html

Ndambuki DK, Al Hitmi HK (2022) Transportation planning and resilience in countries with high risks of natural disasters: the case of Haiti. Soc Sci Res Netw. https://doi.org/10.2139/ssrn.4008746

Ostrom E (2007) Sustainable social-ecological systems: an impossibility? Soc Sci Res Netw https://doi.org/10.2139/ssrn.997834

Palma-Oliveira JM, Linkov I (2017) Resilience and risk: methods and application in environment, cyber and social domains. Springer, the Netherlands. https://books.google.com/books?id=tVovDwAAQBAJ

Palma-Oliveira JM, Trump B (2018) Resilience: moving forward from a metaphor. International risk governance center resource guide. In: Resilience (Volume 2): domains of resilience for complex interconnected systems

Palma-Oliveira JM, Trump B, Wood MD, Linkov I (2018) Community-driven hypothesis testing: a solution for the tragedy of the anticommons. Risk Anal 38:620–634. https://doi.org/10.1111/risa.12860

Pescaroli G, Alexander D (2016) Critical infrastructure, panarchies and the vulnerability paths of cascading disasters. Nat Hazards 82:175–192. https://doi.org/10.1007/s11069-016-2186-3

Pescaroli G, Wicks RT, Giacomello G, Alexander DE (2018) Increasing resilience to cascading events: the M.OR.D.OR. scenario. Saf Sci 110:131–140

Seddon N (2022) Harnessing the potential of nature-based solutions for mitigating and adapting to climate change. Science 376:1410–1416. https://doi.org/10.1126/science.abn9668

Thomas A, Schleussner CF, Kumar M (2018) Small island developing states and 1.5 °C. Reg Environ Change 18:2197–2200. https://doi.org/10.1007/s10113-018-1430-7

United Nations (2015) Transforming our world: the 2030 agenda for sustainable development. NY, USA, New York

Velazquez O, Gianluca P, Gemma C, Carmine G (2020) A review of the technical and socio-organizational components of earthquake early warning systems. Front Earth Sci 8:1–19. https://doi.org/10.3389/feart.2020.533498

Woods Hole (2018) Finding seafloor faults linked to Puerto Rico Quake. Caribbean Tsunami and Earthquake Hazards Studies. Woods Hole Coastal and Marine Science Center, Woods Hole, MA, USA. https://www.usgs.gov/centers/whcmsc/science/caribbean-tsunami-and-earthquake-hazards-studies#:~:text=Earthquakes%20and%20tsunamis%20in%20Puerto%20Rico%2C%20the%20Virgin,the%20rate%20at%20which%20a%20human%20fingernail%20grows

Wu D, Mitchell J, Lambert JH (2022a) Global systemic risk and resilience for novel coronavirus in postpandemic era. Risk Anal 42:1–4

Wu D, Olson D, Lambert JH (2022b) Data analytics and decision-making systems: Implications of the global outbreaks. Decis Support Syst 161:113847–113847

Yew Y, Castro Delgado R, Heslop D, Arcos González P (2019) The Yew disaster severity index: a new tool in disaster metrics. Prehosp Disaster Med 34:8–19. https://doi.org/10.1017/S1049023X18001115

You H, Lambert JH, Clarens AF, McFarlane BJ (2014a) Quantifying the influence of climate change to priorities for infrastructure projects. IEEE Trans Syst Man Cybern: Syst 44:133–145. https://doi.org/10.1109/TSMC.2013.2248709

You H, Connelly EB, Lambert JH, Clarens AF (2014b) Climate and other scenarios disrupt priorities in several management perspectives. Environ Syst Decis 34: 540–554. https://doi.org/10.1007/s10669-014-9525-2

Virgin Islands Established Program to Stimulate Competitive Research (VI-EPSCOR) (n.d.) The National Science Foundation's Established Program to Stimulate Competitive Research (NSF EPSCoR). https://viepscor.org/about-epscor/

Zhuang J, Wang C, Lambert JH (2022) Machine learning and other information analyses for risk in social networks. Risk Anal 42:1603–1605

Chapter 3
Nature-based Solutions for Building Resilience in Small Island Developing States in the Caribbean

Burton C. Suedel, C. D. Metcalfe, Nikolai Emmanuel, and Nealla Frederick

Abstract This article provides some current perspectives on building resilience to climate change in Caribbean Small Island Developing States (SIDS) by describing some examples of the application of Nature-based Solutions (NbS) for protecting coastal and inland areas. In some cases, NbS approaches are integrated into conventional "gray" infrastructure, resulting in hybrid solutions incorporating both natural and hardened structures. We also review the advantages of applying NbS that extend beyond the benefits of conventional approaches. Although some NbS applications can take longer to develop, the benefits can be long-lasting.

3.1 Introduction

Nature-based Solutions (NbS) can be implemented to protect communities and ecosystems from the destructive forces of natural and man-made hazards by utilizing the various services that nature provides. The term "Nature-based Solutions" was first proposed in a report of the Millenium Ecosystem Assessment (2005) and this concept has since been widely adopted by international organizations, government agencies,

B. C. Suedel
US Army Engineer Research and Development Center, Vicksburg, MS, USA
e-mail: burton.suedel@usace.army.mil

C. D. Metcalfe (✉)
The School of the Environment, Trent University, Peterborough, ON, Canada
e-mail: cmetcalfe@trentu.ca

N. Emmanuel
Vetiver Network International, Port of Spain, Trinidad and Tobago
e-mail: bov.iamovement@gmail.com

N. Frederick
The Nature Conservancy Caribbean Programme, St. George's, Grenada
e-mail: nfrederick@tnc.org

C. D. Metcalfe and E. R. Bennett (eds.), *Building Resilience to Climate Change in Small Island Developing States (SIDS) in the Caribbean*, Environmental Contamination Remediation and Management, https://doi.org/10.1007/978-3-031-37376-3_3

researchers, and practitioners (Oen 2019). The International Union for Conservation of Nature (IUCN) defines NbS as "*actions to protect, sustainably manage and restore natural or modified ecosystems that address societal challenges effectively and adaptively, simultaneously providing human well-being and biodiversity benefits*" (Cohen-Shacham et al. 2016). Instead of being viewed as a single solution, NbS should be thought of as a collection of different strategies (Table 3.1).

Climate change is expected to increase both the severity and frequency of extreme weather events in the Caribbean, as well as contributing to a rise in sea levels (Reyer et al. 2017). Small island developing states (SIDS) in the Caribbean are particularly vulnerable to these impacts because of their long coastlines relative to land area and because their populations, which live primarily in coastal areas, cannot escape the effects of extreme weather events by moving far inland. Muñoz and Ötker (2018) described the range of responses that are needed in the Caribbean to build resilience to these events, including "*structural protection to reduce disaster risks*".

Conventional or "gray" engineered structures are an option for building resilience in Caribbean SIDS. However, various NbS approaches are appropriate and in some cases are more cost-effective (Banerjee et al. 2018). In fact, NbS can be viewed within the context of a continuum from gray infrastructure to "green" measures to

Table 3.1 Different concepts and approaches related to NbS. Adapted from Cohen-Shacham et al. (2016)

NbS approach	Sub-category	Examples
Ecosystem restoration approaches	• Ecological restoration • Ecological engineering • Forest landscape restoration	Focus on the recovery of degraded, damaged or destroyed ecosystems
Issue-specific ecosystem-related approaches	• Ecosystem-based adaptation (EbA) • Ecosystem-based mitigation • Climate adaptation services • Ecosystem-based disaster risk reduction (DDR)	Approaches can range from constructed wetlands for flood control to mangrove restoration for DDR
Infrastructure-related approaches	• Natural infrastructure • Green infrastructure	Commonly found in cities. Examples include roof gardens and permeable surfaces
Ecosystem-based management approaches	• Integrated coastal zone management • Integrated water resource management	Both approaches incorporate aspects of conservation and environmental and ecosystem management
Ecosystem protection approaches	• Area-based conservation approaches, including area management	

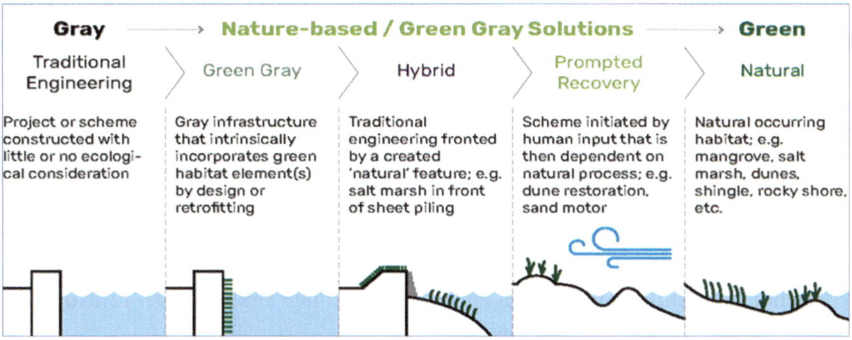

Fig. 3.1 Approaches that incorporate NbS to build resilience can take multiple forms, representing a continuum of measures from gray to green, applied over broad scales and structure types in both coastal and fluvial environments. Coastal systems are illustrated here Figure adapted from Suedel et al. 2021

protect or enhance natural ecosystems (Fig. 3.1). These can be applied over a variety of scales to protect coastal zones and to prevent or mitigate damage from flooding in fluvial environments.

In Caribbean SIDS, NbS strategies for building resilience to climate change and sea level rise offer multiple opportunities under the approaches summarized in Table 3.1. Such strategies can be designed to reduce sediment and nutrient transport, flooding, and landslides in watersheds, increase infiltration to bolster groundwater resources, protect coastal zones, beaches, and infrastructure from storm surge and coastal flooding, sustainably manage fisheries (e.g., Marine Protected Areas), and protect or restore coral reefs. Features integrated into NbS can span a broad range of approaches, whether they be used to enhance conventional built infrastructure or used as part of a hybrid system that includes both gray and green features (Fig. 3.1). This perspectives paper provides a review of NbS approaches for building resilience to natural disasters and sea level rise in Caribbean SIDS. Several NbS projects from the region are presented as case studies to illustrate how some of the challenges to building resilience were addressed. The intent is to offer readers with examples of how these challenges have been overcome at various temporal and spatial scales.

3.2 Case Study: Mitigating Flooding in Trinidad and Tobago

One NbS project located in Trinidad and Tobago focused on mitigating flooding in a residential community located downslope from a large agricultural region. Due to the improper design of the agricultural plots (Fig. 3.2), the farmers did not make use of contour or terrace farming and instead built drainage channels that funneled surface runoff downhill and into the natural valley of the watershed, causing flooding in the community. To mitigate these impacts, a "Vetiver System" was installed by planting the vetiver grass (*Chrysopogon zizanioides*) as hedgerows on the contours (Fig. 3.2). This NbS approach provides several functions that traditional gray infrastructure would not be able to easily facilitate, given the topography of the location. The Vetiver System installed on contours slows the velocity of surface runoff and captures sediment behind the thick leaves of the vetiver hedge, thereby retaining surface soils in the fields. In addition, 22 live stake check dams were constructed using the locally sourced and abundant glory cedar (*Gliricidia sepium*). The living check dams were installed in the gullies of the watershed (Fig. 3.3) to improve the retention of larger sediments and debris that had previously overwhelmed the drainage infrastructure of the residential community below.

Given the steep topography at the site and lack of vehicular traffic access, it would have been costly and time-consuming to engineer a conventional "hardened" drainage management solution to this problem, given that infrastructure logistics such as road access would first need to be constructed to access the site. The ability to access difficult terrain with a relatively small payload of materials allowed for significant cost savings while achieving the desired effect of erosion control. Residents

Fig. 3.2 An example of an NbS approach in Trinidad and Tobago, where vetiver (*Chrysopogon zizanioides*) hedgerows were planted on hillside agricultural property with improper surface water runoff management that was severely eroding the hillside, reducing soil fertility and posing a flooding hazard to downslope residents. The vetiver hedgerows successfully mitigated both hazards *Photo* Nikolai Emmanuel

Fig. 3.3 A live stake check dam made from locally sourced materials was successfully established and is maturing. Hay bales were placed at the base of the living dam to promote the retention of finer-grained sediment. A visible layer of sediment collected by the structure can be observed on top of the hay bales *Photo* Nikolai Emmanuel

in the community indicated to the project sponsor IAMovement (iamovement.org) that a significant reduction in street sedimentation was observed one year after the installation of the vetiver hedgerows and living check dams.

3.3 Case Study: Protecting Coastal Zones in Grenada

An example in Caribbean SIDS which highlights how the benefits of NbS can be achieved over longer time scales is the coastal zone pilot project entitled, "At the Water's Edge" (AWE) in Grenada. The AWE is a highly innovative, unique, multi-year hybrid project being implemented in the Grenville Bay Area of Grenada by The Nature Conservancy (TNC) and project partners that are providing benefits to the coastal communities of Telescope, Grenville, Soubise, and Marquis. It utilizes coastal resilience approaches to demonstrate innovative, cost-effective solutions to address coastal climate impacts on the island. The solutions are scalable and can be replicated by other SIDS contending with similar climate-based challenges. AWE's main goal is to work with coastal communities to enhance their resilience to the impacts of climate change, using ecosystem-based adaptation (EbA; also known as NbS) as alternatives to conventional hardened infrastructure. Building upon the Grenada government's stated strategies for climate adaptation and mitigation, while also addressing specific Sustainable Development Goals (SDGs) as set forth by the United Nations (e.g., SDG 11, SDG 14), AWE interventions are guided by the "Living Edge Plan" (LEP), a series of linked actions with policies to strengthen the resilience of coastal communities. The LEP was developed in collaboration with communities, local and international partners, and government agencies.

In Telescope, the main community in which AWE is being implemented, the fringing reef within the Grenville Bay has deteriorated significantly over the past 25 years due to erosion and other natural and human influences (Fig. 3.4). This is evidenced by a substantial decrease in the reef's crest and a significant reduction in live coral cover, which impedes reef regeneration. As a result, this reef, which is the coastline's first line of defense for reducing wave energy, is unable to function as it should. This has caused a decrease in marine biodiversity and has left residents exposed and vulnerable to storm surges, coastal flooding, and severe coastal erosion (Reguero et al. 2018). These impacts threaten lives, homes, and critical coastal infrastructure.

Utilizing more than 50 years of data for the Grenville Bay Area, AWE is being implemented through two main components; (i) a set of complementary NbS interventions which include stabilizing the shoreline using a "living shoreline" approach involving beach re-nourishment and revegetation and mangrove restoration, and (ii) reef re-engineering which includes coral restoration. Combined, these interventions will provide long-term coastal protection to Telescope's shores. Modeling approaches were used to design and optimize the placement of hybrid structures for reef re-engineering, and in 2015, these structures were placed along a section of the reef

Fig. 3.4 Coastal erosion in Telescope, Grenada in 2014 before NbS was implemented. © The Nature Conservancy

Fig. 3.5 At the Water's Edge innovative pilot reef structures like this one were placed along the shoreline in Telescope, Grenada. The structures were monitored for several years to evaluate their success based on several pre-determined criteria. © Tim Calver for The Nature Conservancy

(Fig. 3.5). This pilot project was initiated to evaluate whether they meet the following criteria: (a) reduce 80–90% of the wave energy passing over them; (b) promote natural biological growth and accretion of coral and crustose coralline algae; (c) be constructed with small boats and local community labor; and (d) be constructed with common local materials at a cost less than that of conventional hardened infrastructure (Reguero et al. 2018).

The pilot projects were monitored for several years, and the lessons learned have been employed in the re-design of the full buildout, which will consist of 300 m of hybrid reef structures which includes coral out-planting along the Grenada shoreline.

There are multiple benefits that are being realized because of the implementation of these NbS approaches. Engineering benefits include the reduction in wave energy and longshore currents, positively impacting local hydrodynamic conditions and thereby reducing erosion and run-up flooding at the beach. As the structures are seeded with corals and become reefs, they will grow and maintain their function, while providing habitat for fish and other vital marine life within the Grenville Bay Area, creating multiple environmental benefits in the process. Socio-economic benefits center around the expected increase in fish populations. With an increase in fish available to local fishermen, this will lead to increased economic activity within the communities, and in turn enhance their food security. A supplemented source of income for the fishermen and community members was also realized when they were hired during the implementation of AWE's full buildout.

Mangroves can recover quickly relative to other marine ecosystems (Beck et al. 2020). This is especially true in situations where inundation by seawater and freshwater hydrology have not been altered by human intervention. Yet, such ideal conditions are often not maintained in areas impacted by coastal development and/or climate change-related hazards. Initial recovery of mangroves after a hurricane has been observed in 3–5 years, although full recovery of a fully functioning mangrove forest can take much longer, depending on the wind velocity, canopy height, dominant tree species, and position of the stand within the forest (Imbert 2018). In a study by Beck et al. (2020), the authors identified multiple cost-effective opportunities for mangrove restoration across the Caribbean. The authors identified 20 locations in the region where the cost–benefit ratio of mangrove restoration exceeded one, and there were many areas where this ratio was exceeded by up to ten times when the mangrove forests were considered as 30-year assets. The AWE also includes a project to restore mangrove ecosystems along vulnerable coastal zones in Grenada.

3.4 Converting Climate Challenges to Opportunities

As discussed in the introductory chapter of this book, SIDS in the Caribbean vary widely in terms of access to economic and human resources, population size, geography, systems of governance, and culture. SIDS in the Caribbean also vary in terms of vulnerability to extreme weather events and sea level rise (Rhiney et al. 2015). However, there are some common features among these nations that contribute to challenges for building resilience by adopting NbS approaches. These barriers include a lack of financing and insufficient or poorly enforced policies for managing development (CCCCC 2012). As identified in other chapters in this book, stakeholder engagement and lack of human resources within local communities are also major impediments to implementing NbS. Yet, evidence both from the authors in this book as well as from other investigators conducting research in SIDS shows that these impediments can be overcome in practice.

Economic impediments include costs related to initial design, construction, and future maintenance of the completed structures. One means of overcoming this challenge is to consider all aspects during the initial analysis of costs and benefits. For example, when applying cost–benefit analysis (CBA) to evaluate NbS projects, it is important that the benefits not be limited to economics but should also incorporate consideration of both the present and future value of all aspects of the project, including environmental and social benefits (Hahn 2021). Environmental benefits that should be considered in the CBA include ecosystem services and ways in which the project improves infrastructure, as well as community resilience. Banerjee et al. (2018) developed a model for estimating the economic benefits and applied it to a shoreline stabilization coastal infrastructure project in Barbados. Their results showed that the project realized meaningful benefits through reduced property damages and erosion to beaches, benefitting both tourists and residents alike.

The extended temporal scales for NbS projects are considered another impediment, as some projects may require longer periods of time to fully realize benefits due to natural infrastructure characteristics, such as the time needed for native plants to sufficiently mature or the relatively slow pace of coral restoration. Local communities may expect immediate benefits from NbS projects, but it should be communicated at the outset that the overall system will provide greater benefits over the longer term (months to years), even if full benefits are not possible over the short or medium term. International guidelines were recently published that acknowledge that NbS may take time to fully achieve objectives to build resilience, and the importance of communicating and managing NbS expectations over the project life cycle (Bridges et al. 2021).

Stakeholder engagement is another area where impediments can be converted to opportunities. Stakeholder engagement should be meaningful in ways that bring together all the stakeholders that are impacted by and benefiting from NbS. One means of accomplishing this is to develop a stakeholder map as the first step when engaging stakeholders. A stakeholder map draws from direct communications with local stakeholders. The map shows individual stakeholders, including individuals, groups, organizations, and decision-makers who may affect, be affected by, or perceive themselves to be affected by NbS project activities. Stakeholder mapping has been successfully used to identify stakeholders in coastal New Jersey, USA, where the resilience of both community and natural infrastructure is of concern in the face of sea level rise and other aspects of climate change (Thorne et al. 2022). Education/outreach materials should also be developed to foster intentional engagement to local communities and decision-makers. Such materials should be framed in terms of information that matters most to those specific individuals and considers the financial status of local communities, their socio-economic issues, and their value and belief systems.

An example of a successful stakeholder engagement process is occurring at Walkers Reserve in Barbados where 112 ha. (277 acres) of an abandoned sand quarry are being converted into a living laboratory for addressing climate resilience. Here, members of the local community are granted access to native coconut trees growing on the reserve. The coconuts are harvested using traditional techniques (Fig. 3.6) and are sold in local markets, thereby contributing to the local economy.

A project conducted in the Fond d'Or watershed on the island of St. Lucia illustrates the importance of considering beliefs when working with critical stakeholders (Clauzel 2011). This watershed is the second largest in St. Lucia, comprising 4,130 hectares (10,230 acres). There are 15 communities within the watershed with 2,760 households, plus 29 scattered settlements.

In some parts of the watershed, the riverbanks are degraded by intensive farming practices and ill-advised grazing of livestock, which leads to soil erosion and flooding of downstream settlements and roads during rain events. A collaborative effort between the private sector, community partners, and farmers led to a riverbank stabilization project along the Grande Ravine tributary of the watershed, which included a reforestation initiative. Selecting the tree species for reforestation was an important consideration for this activity. Ultimately, mahogany (*Swietenia mahagoni*) was

Fig. 3.6 Coconut trees at Walkers Reserve in Barbados. Members of the local community are granted access to coconut trees growing on the property. The coconuts are sustainably harvested and are sold in local markets *Photo* Burton Suedel

selected for the reforestation project. Although this tree species is not native to St. Lucia, it is native to other islands in the Caribbean, and it has high commercial value. However, some farmers were initially reluctant to plant the mahogany saplings due to a perception that these trees contribute to the drying of the soil, and because they do not generate revenue over the short term. Ultimately, the farmers supported the reforestation project once reassured by forestry experts that the layer of litter produced by mature trees conserves soil moisture by slowing evaporation and protects the soil from erosion by absorbing raindrops as they fall.

This case study in St. Lucia also demonstrates that for NbS to meet community as well as ecosystem resilience needs, both scientific and local knowledge about ecosystems should be used to inform NbS implementation. To enable such sustainable management practices, NbS teams consisting of researchers, practitioners, and policymakers are needed to establish and improve an evidence base that draws together scientific, traditional, and experiential knowledge about the scales and contexts in which NbS are effective (Seddon et al. 2020). Such integrated teams are especially needed at the national level, where unique combinations of socio-economic, political, and ecological conditions and belief systems require island-specific approaches, with suitable metrics or indicators to measure progress toward meeting resilience goals. Meaningful integration of researchers, local communities, and traditional ecological knowledge also helps to foster local participation and promotes more equitable and inclusive approaches to NbS (Mercer et al. 2012).

Funding and finance are also important to advance NbS practices. In the case of Caribbean SIDS, limited access to financial resources often must be overcome. One potential means of overcoming this barrier is to integrate NbS with initiatives for promoting green infrastructure, ecosystem restoration, biodiversity, sustainable urban development, natural resources management, climate change mitigation and adaptation, and disaster risk reduction (Cohen-Shacham et al. 2016). These types of large integrated projects can be supported by the Green Climate Fund operated under the auspices of the UN Framework Convention on Climate Change (UNFCCC), which is dedicated to supporting developing countries to mitigate and adapt to climate change. In the Caribbean, the Caribbean Community Climate Change Centre is a "Direct Access Entity" that can broker these types of projects in the region. However, small-scale incentive programs can also be initiated at the local level. In another example of a case study from the Fond D'Or watershed project in St. Lucia, a farmer operating a piggery that was contributing to fecal pollution and soil erosion in the watershed entered into an agreement to mitigate these impacts in exchange for free technical support and labor from local NGOs (Clauzel 2011).

Insurance companies are becoming increasingly aware that they must develop incentive programs to reduce losses due to the impacts of climate change. Taking lessons from usage-based insurance in the automobile insurance business, carriers can incentivize policyholders who invest in mitigating climate-related risks by using a variety of measures such as providing discounts on insurance premiums (Deloitte Center for Financial Services 2022).

Generalized strategies have been developed to foster resilient communities and infrastructure that could be applied in Caribbean SIDS. For example, Steffen et al. (2018) indicated that effective resilience-building strategies should include developing insurance, buffers, redundancy, and diversity, among other initiatives that are needed to transform built infrastructure in the face of climate change. Features of such strategies include (i) maintenance of diversity, modularity, and redundancy; (ii) management of connectivity, openness, slow variables, and feedbacks; (iii) understanding social–ecological systems as complex adaptive systems; (iv) encouraging learning and experimentation; and (v) broadening participation and building trust to promote polycentric governance systems.

3.5 Conclusions

In this chapter, we provided some current perspectives on climate change resilience in Caribbean SIDS. Despite the various challenges to building resilience in the region, there are multiple examples where these barriers are being overcome by the application of NbS. In many of these case studies, NbS approaches are being integrated into existing projects that employ conventional infrastructure, resulting in hybrid designs incorporating aspects of both natural and hardened structures. There are multiple advantages of applying NbS for addressing climate resilience, including providing environmental, social, and economic benefits that extend beyond the benefits of conventional approaches. While some NbS applications can take longer to fully realize their potential, given time, these associated benefits will be long-lasting and wide-ranging. The recent publication of international guidelines for NbS (Bridges et al. 2021) will aid in applying engineering best practices and will reduce the uncertainty associated with their performance. Public acceptance of NbS can be accelerated through meaningful stakeholder engagement, which is critical to project success. Through this process, local communities in Caribbean SIDS can contribute to both the design and the implementation of NbS. Within these communities, NbS will not be accepted until this approach is viewed as a viable solution.

Acknowledgements This article is a product originating from a workshop held in Barbados in June 2022 on Resilience to Climate Change in Small Island Developing States (SIDS) in the Caribbean. We thank the Centre national de la recherche scientifique based in Paris, France, the US Army Corps of Engineers program for Dredging Operations Technical Support (DOTS), and the National Oceanic and Atmospheric Administration of the USA for funding to support the workshop. The authors thank Erin Bennett, Ricardo Marshall, Andrea Miralles-Barboza, Karl Payne, Kim Waddell, and Jehroum Wood for their productive workshop discussions which informed the content of this paper.

References

Banerjee O, Boyle K, Rogers CT, Cumberbatch J, Kanninen B, Lemay M, Schling M (2018) Estimating benefits of investing in resilience of coastal infrastructure in Small Island Developing States: an application to Barbados. Mar Policy 90:78–87

Beck MW, Heck N, Narayan S, Menéndez P, Torres-Ortega S, Losada IJ, Way M, Rogers M, McFarlane-Connelly L (2020) Reducing Caribbean risk: opportunities for cost-effective mangrove restoration and insurance. The Nature Conservancy, Arlington, VA, USA, 56 pp. Available on-line at: https://www-axa-com.cdn.axa-contento-118412.eu/www-axa-com%2Ff83724d2-fcd0-4a41-bde9-e0330a501d07_tnc_mangrove+insurance_final+hi.pdf

Bridges TS, King JK, Simm JD, Beck MW, Collins G, Lodder Q, Mohan RK (eds) (2021) International Guidelines on Natural and Nature-based Features for Flood Risk Management. U.S. Army Engineer Research and Development Center, Vicksburg, MS, USA, Report Number ERDC SR-21-6. Available on-line at: https://doi.org/10.21079/11681/41946

CCCCC (2012) Delivering transformational change 2011–2021: Implementing the CARICOM regional framework for achieving development resilient to climate change. Caribbean Community Climate Change Centre, Belmopan, Belize, 27. Available on-line at: https://cdkn.org/sites/default/files/files/Executive-Summary_final-version_11.05.122.pdf

Clauzel S (2011) Case Study of the GEF-IWCAM Saint Lucia demonstration project: protecting and valuing watershed services and developing management incentives in the fond D'Or watershed area of Saint Lucia, a final report of the global environment facility-funded integrating watershed and coastal areas management (GEF-IWCAM) Project, 48 pp. Available on-line at: https://iweco.org/sites/default/files/2019-03/GEF-IWCAM_Demo_Case_Study_Saint_Lucia_FINAL.pdf

Cohen-Shacham E, Walters G, Janzen C, Maginnis S (eds) (2016) Nature-based solutions to address global societal challenges. International Union for Conservation of Nature (IUCN), Gland, Switzerland, 97 pp. Available on-line at: https://portals.iucn.org/library/sites/library/files/documents/2016-036.pdf

Deloitte Center for Financial Services (2022) Climate risk: regulators sharpen their focus. Available on-line at: https://www2.deloitte.com/us/en/pages/financial-services/articles/insurance-companies-climate-change-risk.html

Hahn RW (2021) Equity in cost-benefit analysis. Science 372:6541

Imbert D (2018) Hurricane disturbance and forest dynamics in East Caribbean mangroves. Ecosphere. 9:e02231.https://doi.org/10.1002/ecs2.2231. Available on-line at: https://doi.org/10.1002/ecs2.2231

Mercer J, Kelman I, Alfthan B, Kurvits T (2012) Ecosystem-based adaptation to climate change in Caribbean small island developing states: Integrating local and external knowledge. Sustainability 4:1908–1932

Millenium Ecosystem Assessment (2005) Ecosystems and human well-being: synthesis. Island Press, Washington, DC, USA, 137 pp. Available on-line at: https://www.millenniumassessment.org/documents/document.356.aspx.pdf

Muñoz S, Ötker I (2018) Building resilience to natural disasters in the Caribbean requires greater preparedness. International Monetary Fund, Washington, D.C., USA, Available on-line at: https://www.imf.org/en/News/Articles/2018/12/07/NA120718-Building-Resilience-to-Natural-Disasters-in-Caribbean-Requires-Greater-Preparedness

Oen AMP (2019) Nature-based Solutions are gaining momentum from international initiatives promoting environmental, social, and economic benefits. Integr Environ Assess Manag 15:830–831

Reguero BG, Beck MW, Agostini VN, Kramer P, Hancock B (2018) Coral reefs for coastal protection: a new methodological approach and engineering case study in Grenada. J Environ Manage 210:146–161

Reyer CPO, Adams S, Albrecht T, Barsch F et al (2017) Climate change impacts in Latin America and the Caribbean and their implications for development. Reg Environ Change 17:1601–1621

Rhiney K (2015) Geographies of Caribbean vulnerability in a changing climate: issues and trends. Geogr Compass 9:97–114

Seddon N, Daniels E, Davis R, Chausson A, Harris R, Hou-Jones X, Huq S, Kapos V, Mace GM, Raza Rizvi A, Reid H, Roe D, Turner B, Wicander S (2020) Global recognition of the importance of nature-based solutions to the impacts of climate change. Glob Sustain 3(e15):1–12

Steffen W, Rockström J, Richardson K, Lenton TM, Folke C, Liverman D, Summerhayes CP, Barnosky AD, Cornell SE, Crucifix M, Jonathan F, Donges JF, Fetzer I, Lade SJ, Scheffer M, Winkelmann R, Schellnhuber HJ (2018) Trajectories of the earth system in the Anthropocene. Proceed Nat Acad Sci 115:8252–8259

Suedel BC, Naylor LA, Meckley T, Cairns C, Bernier J, Morgereth E, Mears W, Piercy CD, Hofstede R (2021) Enhancing structural measures for environmental, social, and engineering benefits. In: Bridges TS, King JK, Simm JD, Beck MW, Collins G, Lodder Q, Mohan RK (eds) International Guidelines on Natural and Nature-based Features for Flood Risk Management. U.S. Army Engineer Research and Development Center, Vicksburg, MS, USA, pp 701–771

Thomas A, Benjamin L (2018) Policies and mechanisms to address climate-induced migration and displacement in Pacific and Caribbean small island developing states. Int J Clim Change Strateg Manage 10:86–104

Thorne S, Kovacs DC, Gailani JZ, Suedel BC (2022) A community engagement framework using mental modeling: the seven mile island innovation lab community engagement pilot—Phase I. U.S. Army Engineer Research and Development Center, Vicksburg, MS, USA. Report Number ERDC TR-22-12. Available on-line at: https://doi.org/10.21079/11681/44983

Chapter 4
Assessment of the Effectiveness of Nature-Based Solutions for Land Restoration in St. Kitts and Nevis

Halla R. Sahely

Abstract This article presents an overview of a project in St. Kitts and Nevis to strengthen institutional capacity, improve policy frameworks, and conduct pilot projects to address land degradation using Nature-based Solutions (NbS). Pilot projects included mitigation strategies in the College Street Ghaut in St. Kitts, as well as restoration of quarry sites and nearby wetlands and protection of coral reefs located on the island of Nevis. The successes and challenges for the project were identified using criteria recently developed by the International Union for the Conservation of Nature (ICUN) for defining a Global Standard for NbS projects.

4.1 Introduction

Nature-based Solutions (NbS) are defined by the International Union for Conservation of Nature (IUCN) as "actions to protect, sustainably manage, and restore natural or modified ecosystems, that address societal challenges effectively and adaptively, simultaneously providing human well-being and biodiversity benefits" (IUCN 2016). NbS are key components of a project in the Caribbean with the title, "Integrating Water, Land and Ecosystems Management (IWEco) in Caribbean Small Island Developing States". St. Kitts and Nevis is one of ten countries participating in this multi-focal regional project with funding from the Global Environment Facility (GEF). The regional office of the United Nations Environment Programme (UNEP) is the lead Implementing Agency, in partnership with The United Nations Development Programme (UNDP). The United Nations Environment Caribbean Regional Coordinating Unit (CAR/RCU) and the Caribbean Public Health Agency (CARPHA) serve as co-Executing Agencies. For project activities in St. Kitts and Nevis, the Department of Environment within the government Ministry of Environment, Climate Action and Constituency Empowerment is the lead implementing agency.

H. R. Sahely (✉)
Independent Environmental Consultant, Basseterre, St. Kitts and Nevis
e-mail: halla@sahely.com

C. D. Metcalfe and E. R. Bennett (eds.), *Building Resilience to Climate Change in Small Island Developing States (SIDS) in the Caribbean*, Environmental Contamination Remediation and Management, https://doi.org/10.1007/978-3-031-37376-3_4

The objective of IWEco is to contribute to the preservation of Caribbean ecosystems and the sustainability of livelihoods through improved fresh and coastal water resources management, sustainable land management, and sustainable forest management. IWEco is an integrative approach which focuses on systems, people and values. All project activities consider whole ecosystems, such as watersheds and focus on relationships and processes within these systems. Notably, this approach puts people and sustainability at the heart of environmental management and ensures intersectoral cooperation and stakeholder engagement. Finally, this approach recognizes that our environment provides us with important and valuable benefits (also called ecosystem services) which support all life and seeks to enhance the sharing of those benefits.

The overall goal of the sub-project in St. Kitts and Nevis is to reduce and reverse land degradation using the IWEco approach. Much of this land degradation is due to extreme weather events, such as tropical storms and hurricanes, as well as drought. St. Kitts and Nevis are within the "hurricane belt" of the eastern Caribbean and past major storms that impacted the region included Hurricane Omar in 2008 (Category 4), Hurricane Earl in 2011 (Category 1), and Hurricane Irma (Category 5) in 2017. Climate change is expected to increase the frequency and the intensity of tropical storms and hurricanes in this part of the Caribbean, and so increasing resilience to these extreme weather events is a priority (Rhiney 2015). Climate change is also expected to increase the frequency and duration of drought in the Caribbean (Herrera et al. 2020), and so measures are also needed to increase resilience to extended drought conditions.

The regional IWEco project in St. Kitts and Nevis focuses on strengthening the institutional capacity, improving the policy framework, and facilitating pilot projects within the College Street Ghaut watershed located on the island of St. Kitts, and key quarry sites and nearby wetlands and coral reefs located on the island of Nevis. Specifically, in St. Kitts, about 1.5 ha. of land located within a major "ghaut" (i.e., ravine) were restored to mitigate against erosion and land degradation. This restoration work in the College Street ghaut included installation of gabion baskets and planting of deep-rooted vetiver grass. In Nevis, restoration activities were initiated over an area of 8 ha. at three separate sites, including an abandoned quarry at Potworks Estate and a coastal wetland at Nelson's Spring. Finally, marine-based NbS were also implemented in the coastal zone of Nevis, which included the installation of artificial reef modules and the establishment of a coral nursery.

4.2 Principles of NbS Adopted by the IUCN

In 2016, the IUCN adopted a concrete definition and goal for NbS and outlined its key guiding principles. The goal of NbS is "to support the achievement of society's development goals and safeguard human well-being in ways that reflect cultural and societal values and enhance the resilience of ecosystems, their capacity for renewal and the provision of services". NbS are designed to address major societal challenges, such as climate change, food security, water security, human health, disaster risk reduction, and social and economic development. The guiding principles for the application of NbS are to.

- embrace nature conservation norms (and principles);
- be implemented alone or in an integrated manner with other solutions to address societal challenges (e.g., with technological and engineering solutions);
- be determined by site-specific natural and cultural contexts that include traditional, local, and scientific knowledge;
- produce societal benefits in a fair and equitable way and in a manner that promotes transparency and broad participation;
- maintain biological and cultural diversity and the ability of ecosystems to evolve over time;
- be applied at a landscape scale;
- recognize and address the trade-offs between the production of a few immediate economic benefits for development, and future options to produce the full range of ecosystems services; and
- be an integral part of the overall design of policies, and measures or actions that address a specific challenge.

Building on the principles outlined above, the IUCN developed in 2020 a "Global Standard" for verification, design, and scaling up of NbS (IUCN 2020). The Global Standard highlights 8 criteria and 28 associated indicators for NbS, addressing the pillars of sustainable development (i.e., biodiversity, economy, and society) and resilient project management (Fig. 4.1). Criteria 1 and 2 speak to identifying the societal challenge to be addressed and designing an appropriate NbS, especially in terms of key spatial considerations. Criteria 3, 4, and 5 correspond to the three key dimensions of sustainable development, which are protecting biodiversity, ensuring economic viability, and promoting social equity. Finally, criteria 6, 7, and 8 address resilient project management and mainstreaming within policy frameworks.

As part of the development of the NbS standard, a self-assessment tool, in the form of an Excel spreadsheet was developed by the IUCN to enable users to identify the extent to which their interventions adhere to the indicators. This chapter will describe the nature of the NbS applied in St. Kitts and Nevis and will demonstrate the application of the self-assessment tool to evaluate the effectiveness of these NbS and to identify the strengths and weaknesses of the pilot projects relative to all eight criteria. This application of the tool is exploratory in nature and should not be interpreted as a strict evaluation of the project interventions.

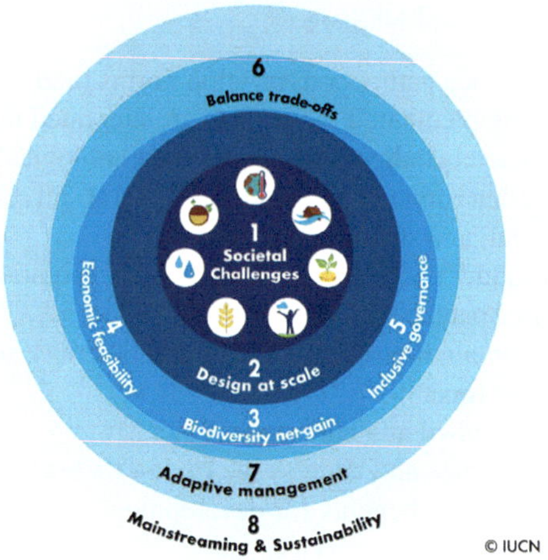

Fig. 4.1 The eight criteria outlined in the IUCN Global Standard for NbS (IUCN 2020)

4.3 Land Degradation Control Works in the College Street Ghaut, St. Kitts

The watershed of the College Street Ghaut covers 662 ha. The watershed has a range of features, including upland natural forest, former sugar cane lands, and urban development centered in the capital city of Basseterre. Land use changes, especially due to urban growth have resulted in rapid surface water runoff and severe soil erosion following rainfall events. The impacts of this land degradation are far-reaching and include public health risks due to flooding and inappropriate disposal of liquid and solid waste, as well as loss of topsoil, poor water quality at the outlet of the ghaut, and contamination of the nearshore marine environment. Major flood events linked to land degradation and chronic sedimentation have been a frequent occurrence, with the last major event in the summer of 2013 that caused an estimated US$120,000 in damages. Works to reduce land degradation inside the ghaut are a critical first step leading to more proactive management of this important zone. The restoration works began in earnest in February 2020 after a clean-up exercise where excess vegetation and solid waste were removed from a 1.5 ha. area targeted by the project, as shown in Fig. 4.2.

The main land degradation control measures implemented were the installation of 325 gabion baskets, repair of two existing weirs, and the planting of vetiver grass. As outlined in Chap. 3 in this book, these hybrid measures utilized both "grey" and "green" infrastructure. A gabion is a galvanized wire cage, cylinder, or box filled with materials like concrete, stones, sand, or soil. When filled with these materials, gabion structures become a cost-effective defense against erosion, and maintain stability and protect streambanks and beds. Historically, gabion baskets were utilized in the

Fig. 4.2 The target location within the College Street Ghaut, St. Kitts, indicated by a circle on a map of the watershed (a) and by a photograph of the site (b)

Federation of St. Kitts and Nevis, especially in the agricultural sector (notably during the times of sugar manufacturing) as a means of soil stabilization along the banks of major ghauts. The works also saw the repair of two weirs that were installed inside the ghaut over 50 years ago (Fig. 4.3). The weirs act as a barrier for sediments and when dredged in a timely manner, they prevent large volumes of sediment from being transported downstream into the coastal zone. Excess debris and sediments were also cleared from culverts and drains, including at the outlet of the ghaut. Over 500 metric tons of sediment were dredged from the outlet of the ghaut in the period from August 2020 to September 2021. A little less than a year later in June 2022, another 350 metric tons of sediments were dredged. This provides some anecdotal evidence that the intervention taken over the past few years are resulting in less sediment being washed away into the coastal waters.

Deep-rooted vetiver grass was also planted along the banks of the ghaut (Fig. 4.4). As described in Chap. 3 in this book, this type of vegetation prevents surface erosion and shallow slides by slowing the velocity of runoff and holding the soil in place through stabilization from the root systems. In addition, this grass, known locally as "khus khus" or "bedgrass", is used for weaving and handicrafts, thus providing a

Fig. 4.3 Gabion baskets installed (a) and weir repair (b) in the College Street Ghaut, St. Kitts

Fig. 4.4 Vetiver grass planted in the College Street Ghaut, St. Kitts, in June 2020 (**a**), after one year (**b**) and after two and a half years of growth (**c**)

livelihood for many persons. Over 500 clumps of grass were planted over the lifetime of the project.

To address the problem of greywater disposal into the College Street Ghaut from adjacent homes, the IWEco project also commissioned the design of a pilot greywater collection, treatment, and reuse systems utilizing constructed wetlands. This work is detailed by Daniel and Paul (2021). It is anticipated that this pilot system will be built with funding from the upcoming GEF-CREW + project in 2023.

Overall, the suite of interventions taken in this section of College Street Ghaut (i.e., a modified ecosystem) is a good example of NbS, as the actions taken significantly reduce the impacts of excessive soil erosion and flooding and increase resilience to the impacts of climate change. The solutions in place now allow for sustainable management of the ghaut. The positive benefits to human well-being include mitigation of flooding risk and associated damage to infrastructure, and improved water quality in receiving waters. Other benefits include the propagation of vetiver grass and other native species which increase biodiversity and prevent soil erosion but can also be used for livelihood generation.

4.4 Reforestation and Restoration at Three Sites in Nevis

A robust reforestation and rehabilitation effort is being spearheaded by the Nevis Historical and Conservation Society (NHCS). This sub-project officially began in May 2020, with a planned duration of 2 years. Building on a previous successful GEF Small Grant Programme project at Coconut Walk Estate (2017–2019), the IWEco project in St. Kitts and Nevis provided a grant of USD 150,000 to the NHCS to continue its reforestation program at Coconut Walk (approximately 2 ha.), to reforest 1.5 ha. of degraded lands at the abandoned quarry at Potworks Estate and to start a pilot wetland restoration effort at Nelson's Spring (approximately 2 ha.).

4.4.1 Coconut Walk Estate

The site at Coconut Walk has seen persistent land degradation over decades due mostly to overgrazing, unsustainable farming practices, and upstream quarry operations. The site is characterized by harsh conditions of a wind-swept landscape with constant sea blast, rocky and arid topography, degradation of the original vegetation, diminished rainfall in recent years and lack of water retention capacity. Despite this challenging environment, the NHCS has adopted several interventions aimed at sustainable reforestation of this site. The entire site has been fully contoured with berms to restrict erosion, under which swales were developed for water retention and with biomass to support vegetation regrowth. Additionally, sediment traps and small catchment areas were constructed to aid water retention and vegetation regrowth, including cactus gardens. There is now evidence of teeming life of plants, insects, lizards, and birds, with some of the plants established as naturally occurring pioneering species and others propagated (Fig. 4.5). Small-scale composting was developed at the site to promote soil nutrient health. Finally, fencing and electrical wire were erected as deterrents to ruminants, particularly goats and donkeys that previously roamed and grazed in the area.

After much experimentation and consistent effort, native sea grape, coconut, almond, and vetiver grass, including a variety of forest species and fruit trees have adapted and established under the prevailing harsh conditions of strong winds and sea blast, drought, and otherwise barren soils (Fig. 4.6). Natural precipitation was supplemented by drip irrigation from municipal water sources, when available. The area is only sporadically serviced by the municipal water supply during drought conditions. As such, the work crew devised a system to water the newly planted seedlings consisting of a recycled 3.8 L plastic water bottle, perforated at the base, which is placed beside the seedling. The perforated base allows for slow infiltration of water into the root zone (Fig. 4.6). The bottles are refilled manually once a week. Eventually, even this way of irrigation proved not effective, as the perforations

Fig. 4.5 Soil retention techniques at Coconut Walk, Nevis, including construction of contours (**a**) and sediment traps (**b**)

Fig. 4.6 Planted coconut trees (a) and recycled plastic bottle watering system for seedlings (b) at Coconut Walk, Nevis

clogged, and the bottle required filling quite often during the dry season. Additional irrigation stations were set up which consisted of recycled 170 L (i.e., 45-gallon) drums or other large containers.

Overall, there is evidence of progressive vegetation growth and habitation of insects and land crabs in the area, particularly along the berms and swales. Over 2,000 trees were planted during the sub-project at Coconut Walk, but only 731 plants have survived, which amounts to a survival rate of 36%. Many mitigating circumstances contributed to this low rate of survival, including extended drought, inconsistencies in the public water supply, animal damage due to lack of forage foods nearby, high and continuous salt-laden winds, and very poor soils.

4.4.2 Abandoned Quarry at Potworks Estate

Quarrying operations ceased at Potworks Estate over 10 years ago. No restoration efforts were undertaken at the end of the operation, leaving a denuded and exposed hillside prone to severe erosion after heavy rainfall. At the inception of the IWEco project, the site still had abandoned equipment littering the landscape and had become a hotspot for illegal dumping of waste. There was significant evidence of soil erosion and land degradation, with a proliferation of invasive plant species.

Despite the ongoing challenges due to the COVID-19 pandemic, a robust effort of solid waste clean-up and land preparation was undertaken by the NHCS in 2021. Through interventions similar to the site at Coconut Walk, the land was contoured, and berms, sediment traps and catchments were created to restrict stormwater runoff and promote infiltration (Fig. 4.7). Invasive plant species were also removed.

A total of 1,039 plants were planted, and at the end of the project, 490 plants survived, for a rate of survival of 47.2%. Many circumstances contributed to this

Fig. 4.7 Abandoned quarry at Potworks Estate, Nevis, before clean-up (**a**), after land preparation work (**b**), and after one year of active planting (**c**)

relatively low survival percentage, including an extended drought (with this area of the island always seeing less rainfall than other areas), no access to the public water supply, damage from free roaming animals especially donkeys, landslides where the historical quarry operations left steep walls, and very poor soil fertility. More resources would have been required to be able to increase the soil nutrients throughout the site.

4.4.3 Wetland Restoration at Nelson's Spring

The area known as Nelson's Spring occupies approximately 2 ha. on the west coast of Nevis. Within the site is a pond, maintained year-round by the main west coast aquifer and also fed, mainly in the rainy season, by drainage from the slopes of Nevis Peak located to the east. There is one drainage channel to the sea, which is usually blocked by a sand berm that may be breached after heavy rains.

In recent years, Nelson's Spring has been subject to many negative environmental pressures, as neighboring developments have cut off the natural links with other wetlands. There has also been wholesale clearing of vegetation aimed at "cleaning up", without any attempt at replanting. The result of these disturbances has been the invasion of the whole area with non-native species. Around the drier edges, the most frequent colonizing species are casha *(Acacia macracantha)*, clammy cherry *(Cordia obliqua)*, and wild tamarind *(Leucaena leucocephala)*. This has resulted in dense

Fig. 4.8 Arial view of the wetland at Nelson's Spring, Nevis, showing evidence of emergent scrub and cattails throughout the pond

almost impenetrable scrub suppressing the regeneration of native species, including white mangrove *(Laguncularia racemosa)* and seagrape *(Coccoloba uvifera)*. The pond itself has been colonized by southern cattail *(Typha domingensis)* and these fast-growing reeds have displaced native reeds and other species, presenting a threat to the natural biodiversity (Fig. 4.8). They have also severely reduced the open water of the pond and changed the open aspect from the road to the sea (NHCS 2021).

The NHCS has teamed up with a local community group, the St. Thomas' Improvement Group (STIG) to restore the wetland at Nelson's Spring by removing invasive cattail reeds and scrub, thus reopening the beautiful vista to the sea. Clearing began in late November 2020 and proceeded rapidly, as cattails were cleared from the whole perimeter of the pond and scrub was cleared in the southern section to around 15 m out from the pond edge. Replanting began in December 2020. Some surviving coconuts were relocated and red mangrove (*Rhizophora mangle*) were planted near the mouth of the pond. In January 2021, community days were held to plant seagrape, coconut, white mangrove, seaside almond (*Terminalia catappa*), swamp fern (*Acrostichum danaeifolium*), and beach morning glory (*Ipomoea pes-caprae*). Volunteers also helped remove regrowing cattails. Along with the replanting, there was encouraging regrowth of native species of sedge and spike reed (*Eleocharis mutata*). There had been concern for the native spike reed which had once been widely present in the wetland, but which was no longer in evidence among the dense cattails. However, this native plant was soon seen regrowing in cleared areas (Fig. 4.9). By April 2021, the water lily (*Nymphaea ampla)* had re-established and has now spread through the whole pond. Currently, work has centered on slowing down the regrowth of cattails by employing one or more laborers to physically remove plants and cut off flower stalks. A recent purchase of a reed cutter will enable faster clearing (NHCS 2021).

Water birds have been very visible during all stages of the project. In April 2021, a group of young birders on an ornithology course visited Nelson's Spring on a field trip. The local ornithologists accompanying them identified 33 species of birds

Fig. 4.9 Wetland and pond at Nelson's Spring, Nevis after initial clearing and replanting (**a**). Spike reeds (**b**) and sedges (**c**) return to the wetland

in the two-hour visit, more than at any other site they had visited in both St. Kitts and Nevis. Fish were seen jumping in the pond, and land crab holes were evident. Also, recently many butterflies have been seen in the areas where native plants are re-establishing. In addition, the neighboring beach is a nesting site for leatherback turtles (*Dermochelys coriacea*) and hawksbill turtles (*Eretmochelys imbricata*).

In mid-2021, the NHCS applied for and received a grant from the US National Forest Service under their Natural Infrastructure for Caribbean Resilience (NiCAR) program and the Small Island Organization (SMILO) Islands Fund to continue the restoration effort at Nelson's Spring. By 2022, it was clear that the cattails were growing more rapidly than the native species. The STIG has redoubled its efforts and the hope is that with the additional funding sources, a more consistent clearing of the reeds can take place with specialized equipment.

4.5 Coral Reef Monitoring and Restoration

The IWEco project in St. Kitts and Nevis also sought to implement marine-based NbS to help mitigate against the negative impacts of land-based sources of pollution (i.e., LBS). Although a smaller component than the interventions taken on the land, the activities focused on increasing capacity for coral reef monitoring, as well as restoration activities such as artificial reefs and coral nurseries. These activities included: (1) a survey of selected coral reef using a drop-down camera technique, (2) establishment of a coral nursery, (3) training in coral reef monitoring, restoration, and benthic analysis, and (4) installation of the "MoReef" artificial reef modules at three locations in the Federation. Training activities were wide ranging and saw capacity building for over 150 persons in coral reef monitoring and restoration.

Fig. 4.10 MoReefs being transported (**a**) and installation of a MoReef stack (**b**)

One such training program related to coral reef restoration was delivered in May 2021 over four sessions (each session was 2 h) using the Zoom platform. The sessions were facilitated by Dr. Alwin Hylkema of the Van Hall Larenstein University of Applied Science in the Netherlands and based in the Caribbean island of Saba. Dr. Hylkema is also the developer of the "MoReef" artificial reef system. Over 30 participants attended the sessions which covered 4 main topics: (i) artificial reefs, (ii) coral reef restoration, (iii) *Diadema* (long-spined sea urchin) restoration, and (iv) application of the MoReef. Subsequently, the IWEco SKN project together with the Department of Marine Resources moved to construct and deploy 12 MoReef modules at three different locations in the Federation at Church Reef (Nevis), Calypso Reef (St. Kitts) and Green Point (St. Kitts). These were installed in June 2022 and monitored closely for the first four months (Fig. 4.10). Although it is still too early to fully quantify the impact of the artificial reefs, initial surveys demonstrate that marine biodiversity is on the rise in and around the MoReef modules.

In June 2022, an NGO based in Nevis named C.A.R.E. SKN (Coral Assistance, Restoration and Education, St. Kitts and Nevis) was engaged by IWEco SKN to build and establish the first ever coral nursery in the Federation (www.facebook.com/caresaintkittsnevis/). Three "coral trees" were installed at Tamarind Cove, Nevis on July 15, 2022. After coral fragments were collected, founders and co-directors of C.A.R.E SKN, Valérie Grégoire and Ruddell Williams added coral fragments to the coral tree on July 28, 2022 (Fig. 4.11). The focus of this nursery is on the branching staghorn corals, also known as *Acropora sp.* that have been in decline in the Caribbean since the 1970s. These types of branching coral can grow quickly in the right environment. Monitoring is ongoing and early indications show a low mortality rate (less than 20%) and that many of the coral fragments are indeed growing. This pilot program has received a lot of attention and the group has now applied for more funding to continue.

Fig. 4.11 C.A.R.E SKN diver Valérie Grégoire with a fragment of *Acropora palmata* (**a**) and diver Rudell Williams (**b**) attaching fragments to coral trees at Tamarind Cove, Nevis. Photographs provided by CARE SKN

4.6 Application of the IUCN Designed Self-assessment Tool for NbS

The IWEco project in St. Kitts and Nevis lasted about three years and was implemented during the COVID-19 pandemic. Many of the interventions taken clearly fit the definition of NbS, although during the project design phase these interventions were not clearly identified as such but more broadly categorized as relating to land and ecosystems restoration. Even though it is clear that more time is needed to assess the long-term impacts of these interventions, it is useful and insightful to apply the self-assessment tool as outlined in the IUCN Global Standard for NbS to evaluate the effectiveness of these interventions and to highlight lessons learned to inform other similar projects designed and implemented in Caribbean SIDS. The application of the tool in this article should be viewed as representing only the first steps in the process of evaluating the design and implementation of NbS in St. Kitts and Nevis. The effectiveness of these NbS projects are evaluated against 8 criteria using the 28 indicator metrics identified by the IUCN described below:

Criterion 1: NbS effectively address societal challenges

Indicator 1.1: The most pressing societal challenge(s) for rights-holders and beneficiaries are prioritized.

Indicator 1.2: The societal challenge(s) addressed are clearly understood and documented.

Indicator 1.3: Human well-being outcomes arising from the NbS are identified, benchmarked, and periodically assessed.

The IUCN (2020) identified 7 major challenges which NbS should address including: (1) Climate change mitigation and adaptation, (2) Disaster risk reduction, (3) Economic, and social development, (4) Human health, (5) Food security, (6) Water security, and (7) Environmental degradation and biodiversity loss. The IWEco project, which is funded by the GEF, is a multi-focal project with sub-projects throughout the participating Caribbean SIDS. In St. Kitts and Nevis, the main societal challenge being addressed is environmental degradation and biodiversity loss, with some focus on the challenges of disaster risk reduction (DRR) and water security. Figure 4.12 highlights the GEF focal areas which are addressed in the St. Kitts and Nevis sub-project, and we can see these align well with the societal challenges as outlined by IUCN. The project design phase was thorough and lasted several years and saw a wide range of stakeholders consulted both locally and within the wider Caribbean region. As for Indicator 1.3, human well-being outcomes related to public health through improved water quality monitoring and storage were identified as wider goals of the NbS but were not thoroughly benchmarked. They have been periodically assessed as part of the mid-term review of the project done by the GEF. Additionally, a monitoring and evaluation framework established for the project outlines various performance indicators that have been assessed at midterm and are currently being assessed now that the project is near completion. As such, indicators 1.1, 1.2, and 1.3 are adequately addressed and well documented in the IWEco project document (IWEco 2018) and through the mid-term review evaluation (IWEco 2020).

Criterion 2: design of NbS is informed by scale

Indicator 2.1: The design of the NbS recognizes and responds to interactions between the economy, society, and ecosystems.

Indicator 2.2: The design of the NbS is integrated with other complementary interventions and seeks synergies across those sectors.

Indicator 2.3: The design of the NbS incorporates risk identification and risk management beyond the intervention site.

As noted earlier, the design phase of the IWEco project was thorough and took several years. As it relates to small islands and areas with a small physical landmass, the ecological processes across the entire land and seascape need to be considered as

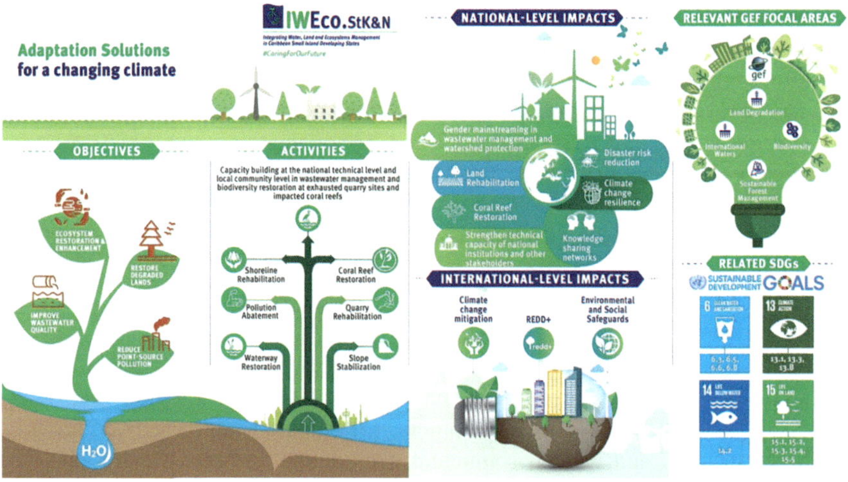

Fig. 4.12 Societal challenges identified and addressed by the NbS interventions during the IWEco SKN project

they are so closely intertwined. For example, in the case of the College Street Ghaut, although the intervention site is only a small part of the entire watershed, a diagnostic analysis/vulnerability assessment conducted before the intervention started identified some of the interactions between the economy, society, and the ghaut ecosystem (Sahely 2019a). As such, Indicator 2.1 is adequately fulfilled. Moving on to indicators 2.2 and 2.3, the project document speaks to integration with other similar projects but given the amount of time that elapsed between design and implementation, synergies across sectors were not fully realized. As to risk identification and management beyond the intervention site, this was considered to some extent in the vulnerability assessment, but only partially. There is no doubt that is it difficult to balance the outcomes and resources allocated to a very small-scale intervention, such as the case of the College Street Ghaut, with the need to consider wider interactions beyond the intervention site. This speaks to the need to allocate serious resources (both human and financial) to this type of whole systems analysis before, during, and after an NbS intervention is taken. Overall, for NbS on small island developing states, this criterion encourages us to expand our boundaries and consider the whole island system if the NbS is truly to be effective and sustainable.

Criterion 3: NbS results in a net gain to biodiversity and ecosystem integrity

Indicator 3.1: The NbS actions directly respond to evidence-based assessment of the current state of the ecosystem and prevailing drivers of degradation and loss.

Indicator 3.2: Clear and measurable biodiversity conservation outcomes are identified, benchmarked, and periodically assessed.

Indicator 3.3: Monitoring includes periodic assessments of unintended adverse consequences on nature arising from the NbS.

Indicator 3.4 Opportunities to enhance ecosystem integrity and connectivity are identified and incorporated into the NbS strategy.

Indicators 3.1 and 3.2 were achieved for this project and can be highlighted by the reforestation and wetland restoration work that took place at three sites in Nevis. At all three sites, there was substantial knowledge about the current state of the ecosystem and the prevailing drivers of degradation which are well documented by the Nevis Historical and Conservation Society in their proposal and project reports (NHCS 2021, 2022). Clear conservation goals were identified, benchmarked, and assessed, which is evidenced by the data collected on types of plants, survival rates, and mitigating circumstances (NHCS 2022). Indicator 3.3 is not as relevant as there were no perceived unintended adverse consequences, although a thorough assessment of such was not conducted. As for indicator 3.4, opportunities to enhance ecosystem integrity are highlighted especially as lessons were learned and key success factors were identified, which included regular access to water for irrigation, fencing to deter feral animals, and need for enhanced soil fertility, among others. Unfortunately, opportunities to enhance ecosystem connectivity were not identified as this would likely outside of the scope of the project and would need more resources to fully flesh out and analyze. As with criterion 2, we can see that in the case of small islands, the whole island system should be considered in the application of any NbS.

Criterion 4: NbS are economically viable

Indicator 4.1: The direct and indirect benefits and costs associated with the NbS, who pays and who benefits are identified and documented.

Indicator 4.2: A cost-effectiveness study is provided to support the choice of NbS including the likely impact of any relevant regulations and subsidies.

Indicator 4.3: The effectiveness of the NbS design is justified against available alternative solutions, taking into account any associated externalities.

Indicator 4.4: NbS design considers a portfolio of resourcing options such as market-based, public sector, voluntary commitments, and actions to support regulatory compliance.

The analysis of costs and benefits, which include both financial and non-financial elements, was identified in the project document and discussed in stakeholder consultations during project design and inception. However, there was not a strong understanding of how the costs and benefits were distributed and this aspect lacks verification. As such, indicator 4.1 was only partially met. As for indicators 4.2 through 4.4, it is obvious that the solutions implemented under the project were less costly than their hard infrastructure counterparts, especially as related to drainage projects. Sahely (2019a) highlighted some of the costs of traditional infrastructure for drainage and flood management in the College Street Ghaut and these are much higher than their green infrastructure equivalents such as the installation of gabion baskets. There was no attempt to estimate an internal rate of return, nor a meaningful review of the NbS cost effectiveness against other viable alternatives. Indicators 4.2 and 4.3 were thus insufficiently met. Finally, if we take the College Street Ghaut as an example, the government of St. Kitts and Nevis has been identified as the long-term funder beyond this piloting phase, but more analysis would be required to properly assess feasibility and replicability in other sections of the ghaut. Indicator 4.4 was thus partially met.

Criterion 5: NbS are based on inclusive, transparent, and empowering governance processes

Indicator 5.1: A defined and fully agreed upon feedback and grievance resolution mechanism is available to all stakeholders before an NbS intervention is initiated.

Indicator 5.2: Participation is based on mutual respect and equality, regardless of gender, age or social status, and upholds the right of Indigenous Peoples to Free, Prior and Informed Consent (FPIC).

Indicator 5.3: Stakeholders who are directly and indirectly affected by the NbS have been identified and involved in all processes of the NbS intervention.

Indicator 5.4: Decision-making processes document and respond to the rights and interests of all participating and affected stakeholders.

Indicator 5.5: Where the scale of the NbS extends beyond jurisdictional boundaries, mechanisms are established to enable joint decision-making of the stakeholders in the affected jurisdictions.

Environmental projects such as this one implemented in St. Kitts and Nevis are almost exclusively implemented and managed by the Government of St. Kitts and Nevis. As for other government processes, there is a defined feedback and grievance resolution mechanism which is available to all residents on the islands. However, the resolution mechanism was not fully vetted nor customized to ensure it was compatible with the IWEco SKN project and to the NbS projects that were implemented. As such, indicator 5.1 is only partially met. The governance process that was put in place at project inception was the establishment of a project steering committee (PSC) which was inclusive and transparent and included stakeholders from all the key groups (including NGOs) which were thoroughly mapped in the original project document (IWEco 2018). The PSC met formally at least once yearly, and the project manager was in constant communication with members throughout the project lifetime. All outputs were shared with the PSC and comments from all stakeholders incorporated. Private citizens residing adjacent to intervention sites were engaged and kept abreast of the activities being undertaken. To achieve more inclusivity, members of civil society and the private sector should have had representation on the PSC. Based on this structure, indicators 5.2 through 5.5 were adequately met.

Criterion 6: NbS equitably balance trade-offs between achievement of their primary goal(s) and the continued provision of multiple benefits

Indicator 6.1: The potential costs and benefits of associated trade-offs of the NbS interventions are explicitly acknowledged and inform safeguards and any appropriate corrective actions.

Indicator 6.2: The rights, usage of and access to land and resources, along with the responsibilities of different stakeholders, are acknowledged and respected.

Indicator 6.3: The established safeguards are periodically reviewed to ensure that mutually agreed trade-offs limits are respected and do not destabilize the entire NbS.

In line with the analysis for criterion 4, insufficient work was done to fully quantify the costs and benefits and associated trade-offs and safeguards at the NbS sites and the larger landscape (i.e., Indicators 6.1 and 6.2). However, as noted in the analysis of criterion 5, stakeholders were mapped adequately and the rights, usage and access to land and resources documented during design, inception and implementation of the project.

Criterion 7: NbS are managed adaptively, based on evidence

Indicator 7.1: A NbS strategy is established and used as a basis for regular monitoring and evaluation of the intervention.

Indicator 7.2: A monitoring and evaluation plan is developed and implemented throughout the intervention lifecycle.

Indicator 7.3: A framework for iterative learning that enables adaptive management is applied throughout the intervention lifecycle.

As part of the project design and implementation, a clear framework was developed which states intended outcomes, actions, and assumptions. The plan was utilized to inform monitoring and evaluation with a series of performance indicators attached to each outcome. All GEF funded projects are set up in a similar manner which is fully fleshed out in the project document (IWEco 2018). In the case of St. Kitts and Nevis, this plan was further refined since several years had elapsed between project design and implementation. Furthermore, during project implementation, the COVID-19 pandemic prompted the PSC to continue to assess the project outcomes and adjust as needed especially as it related to tapping into local entities and knowledge as opposed to international consultants which was impossible during the pandemic. Most of the adaptative management occurred in an organic way and grew out of consistent discussion with key stakeholders and the project steering committee. This type of flexibility and iterative review of the effectiveness of the interventions became a new way of "doing things" and helped keep the project moving forward during the pandemic. Clearly and in hindsight this is a major lesson learned from this project. Undoubtedly, this requires effective and continuous communication between the project management team and its key stakeholders. A project steering committee must be empowered to make "course adjustments" without too many impediments or administrative barriers as long as the overarching objectives outlined under criterion 1 are still being met. Overall, more effort in articulating a well-defined "learning" framework is invaluable especially when unforeseen forces impact project implementation.

Criterion 8: NbS are sustainable and mainstreamed within an appropriate jurisdictional context

Indicator 8.1: The NbS design, implementation, and lessons learnt are shared to trigger transformative change.

Indicator 8.2: The NbS informs and enhances facilitating policy and regulation frameworks to support its uptake and mainstreaming.

Indicator 8.3: Where relevant, the NbS contributes to national and global targets for human well-being, climate change, biodiversity, and human rights, including the UN Declaration on the Rights of Indigenous Peoples (UNDRIP).

The outputs and outcomes of activities and solutions implemented under IWEco SKN have been shared widely, both in the country and regionally using various platforms, including features in the quarterly regional IWEco newsletter, production of several videos and infographics shared locally via the government information service and social media pages of participating NGOs and government departments and presentations at regional webinars and conferences (Sahely 2019b, 2021 and 2022) and nicely highlighted on the project website (https://www.iweco.org/countries/saint-kitts-and-nevis). As the IWEco project wraps up, the regional project is producing a case study of the IWEco SKN sub-project as the primary tool for sharing of lessons learned. As such, indicator 8.1 was adequately met. To bring it up to the next level, a more targeted communications strategy and additional funding allocated to communications could have helped to trigger transformational change. As it relates to activities on the ground for facilitating policy and regulatory frameworks, the IWEco SKN project took proactive measures to aid in the upgrade of the National Conservation and Environmental Management Act of SKN and it is now being considered by national stakeholders with the view of having the amendments presented and passed by the national parliament early in 2023. Finally, as noted in Fig. 4.11, the local NbS contributed to national and global targets specifically SDG 6, 13, 14, and 15. As it speaks to replicability and sustainability, two NGOs in Nevis, the NHCS and C.A.R.E SKN were able to secure additional funding to continue the rehabilitation work initiated under the IWEco SKN project. In addition, the government of St. Kitts and Nevis has in hand essential information as it related to the cost of NbS interventions for ghaut rehabilitation, which can be utilized to upscale and replicate these interventions in the upper reaches of the College Street Ghaut and in other ghauts around the island.

4.7 Concluding Remarks

The IUCN NbS self-assessment tool gives the user a “traffic light” system for each criterion, thereby allowing the user to identify areas for improvement and giving an overall impression of whether the interventions adhere to the IUCN Global Standard. The Global Standard (IUCN 2020) states “for each Indicator, a score out of four is recorded, depending on whether the intervention addresses the indicator to a strong, adequate, partial or insufficient extent. The result is used to calculate the level of adherence to each individual criterion, also giving a strong, adequate, partial and insufficient result for scores greater than 75, between 50 and 75, between 25 and 50 and lower than 25 per cent respectively. These Indicator scores are then normalized so that each Criterion has equal weight. After normalization, the criterion scores are combined to give the overall percentage match. Regardless of the overall percentage match, if an intervention scores an “insufficient” rating against any criterion, then it does not adhere to the IUCN Global Standard for NbS. The percentage match can then be used to describe if this adherence is strong, adequate or partial”. Figure 4.13 displays an overview of the self-assessment exercise for the IWEco SKN project using the scoring system described above which is built into an Excel spreadsheet. Clearly, an oversight of the NbS undertaken in St. Kitts and Nevis relates to economic feasibility and the balancing of the trade-offs which are interlinked. Although it may seem implicit to stakeholders and project designers that the NbS proposed are more cost-effective than traditional hard infrastructure (“gray”) solutions, it is important to be able to quantify this with a thorough cost–benefit analysis. This type of analysis would be critical, especially as we move to replicate and upscale, moving beyond just pilot NbS projects in Caribbean SIDS. Another important weakness is the lack of applying whole systems thinking across the entire island landscape, which is the appropriate scale for such small island nations. Evidently, this would require additional resources both human and financial, but is the best course of action if we are to develop and implement successful NbS in Caribbean SIDS.

Although the IUCN Global Standard for NbS requires adherence to all eight of the criteria, in this exploratory analysis, we can see the strengths of the IWEco SKN projects across criteria 1,7, and 8, which should be replicated and enhanced in nature-based restoration works across the Caribbean. The COVID-19 pandemic highlighted the need for resilient project management and the ability to adapt when faced with uncertainty. The case of NbS in St. Kitts and Nevis demonstrates that true adaptive management requires an empowered project management team and steering committee which communicates consistently with its key stakeholders and partners, including project funders and the executing and implementing agencies. The IUCN Global Standard is an easy-to-use tool to help NbS professionals design and implement effective NbS to address society’s pressing challenges. There is no doubt that NbS are a key element to making “peace with Nature” and truly transforming into “Generation Restoration”.

NbS self-assessment overview

Criterion	Your Criterion Score	Maximum Criterion Score	Normalised criterion	FINAL OUTPUT Your Criterion %age
1. Societal challenges	6	9	0.67	67%
2. Design at scale	4	9	0.44	44%
3. Biodiversity net-gain	6	12	0.50	50%
4. Economic feasibility	2	12	0.17	17%
5. Inclusive governance	9	15	0.60	60%
6. Balance trade-offs	2	9	0.22	22%
7. Adaptive management	6	9	0.67	67%
8. Sustainability and mainstreaming	6	9	0.67	67%
Total Percentage match				49%
Is this in adherance with the IUCN Global Standard for NbS?			Not in adherance	

Key	Output
Strong	Intevention adheres to the IUCN Global Standard for NbS.
Adequate	
Partial	
Insufficient	Intervention does not adhere to the IUCN Global Standard for NbS.

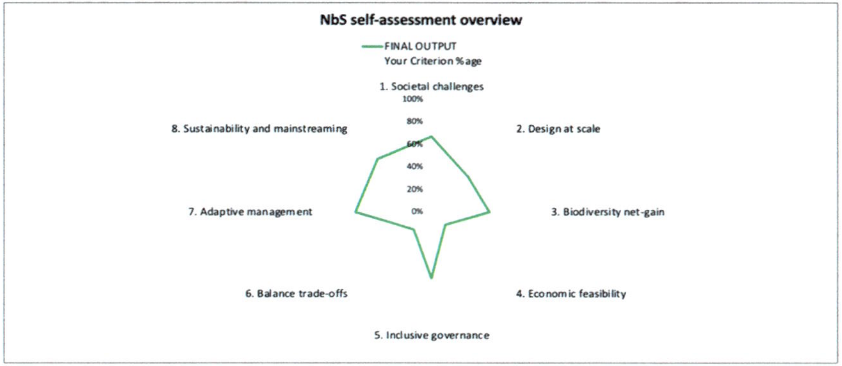

Fig. 4.13 NbS self-assessment overview as applied to the activities of the IWEco SKN

Acknowledgements The author wishes to acknowledge the U.S. Army Corps of Engineers, National Oceanic and Atmospheric Agency, Trent University, and others for the opportunity to participate in the risk forum entitled "Resilience to Climate Change in Small Island Developing States in the Caribbean Through Integration of Engineered and Nature-Based Solutions" held in Barbados in June 2022. Thank you to Dr. Chris Metcalfe, Dr. Erin Bennet, Dr. Adrian Cashman, and Ms. Donna Sue Spencer for providing valuable comments to this paper. The author would like to thank the office of the National Focal Point for IWEco in St. Kitts and Nevis including Ms. June Hughes and Mr. Eavin Parry along with all project partners and agencies. Finally, thank you for donor funding provided by the Global Environment Facility and the hard work of the Regional IWEco Project Coordination Unit and the UNEP Cartagena Convention Secretariat.

References

Daniel A, Paul D (2021) Optioneering and design of small-scale pilot greywater collection, treatment and reuse system, College Street Ghaut, St. Kitts. In: Caribbean water and wastewater association annual conference, Cayman Islands

Herrera DA, Ault TR, Carrillo CM et al (2020) Dynamical characteristics of drought in the Caribbean from observations and simulations. J Clim 33:10773–10797

IUCN (2016) Defining nature based solutions. In: International union for conservation of nature resolution adopted at world conservation congress, Hawaii. Last accessed online on September 17, 2021 at: https://portals.iucn.org/library/node/46486

IUCN (2020) Global Standard for Nature-based Solutions. International Union for Conservation of Nature. Last accessed online on November 14, 2022 at: https://portals.iucn.org/library/sites/library/files/documents/2020-020-En.pdf

IWEco (2018) IWEco SKN National Project Document. Last accessed online on November 14, 2022 at: https://www.iweco.org/sites/default/files/2019-03/IWEco_Project%20Document_StKitts_Nevis_2018.pdf

IWEco (2020) Midterm Review of the UNEP/GEF IWEco Project. Last accessed online at: https://www.iweco.org/sites/default/files/2021-05/IWEco_MTR_Report_FINAL_Sept2020_v2.pdf

Nevis Historical and Conservation Society (NHCS) (2021) Restoration of Nevis Wetlands: Nelson's Spring and Fort Ashby. Proposal submitted for grant funding to the US Forest Service, Natural Infrastructure for Caribbean Resilience (NiCAR) programme

Nevis Historical and Conservation Society (NHCS) (2022) Final terminal report for IWEco pilot project titled Reducing land degradation at Hicks/Potworks Estate, New River Estate, and Nelson's Spring, Nevis via sustainable land management, reforestation, and wetland restoration

Rhiney K (2015) Geographies of Caribbean vulnerability in a changing climate: issues and trends. Geogr Compass 9(3):97–114

Sahely HR (2022) Catalyzing adaptive management in the implementation of nature-based solutions in St. Kitts and Nevis. In: Caribbean water and wastewater association annual conference. Nassau, Bahamas

Sahely HR (2021) Making Peace with Nature—Using the IWEco approach to ecosystems restoration in St. Kitts and Nevis. In: Caribbean water and wastewater association annual conference, Cayman Islands

Sahely HR (2019a) IWEco SKN diagnostic analysis: college street ghaut. Report prepared for the IWEco SKN project

Sahely HR (2019b) Integrating Water, Land and Ecosystems Management (IWEco) as a means of enhancing climate resilience: The case of St. Kitts and Nevis. Caribbean Water and Wastewater Association Annual Conference, Basseterre, St. Kitts

Chapter 5
Building Resilience for Caribbean Nations: Showstoppers and Opportunities from Economics and Governance

Stephanie E. Galaitsi, Christopher Corbin, Jesse Keenan, Shelly-Ann Cox, Genora Joseph, Patrick McConney, Adrian Cashman, Cletus Springer, Christopher Cummings, Benjamin D. Trump, and Igor Linkov

Abstract Small Island Developing States (SIDS) in the Caribbean contribute minimally to the global emissions of greenhouse gases that are driving climate change. However, Caribbean SIDS are disproportionately affected by climate change, as well as impacts from other natural hazards such as volcanic eruptions and tsunamis. Historically, programs and investments in the Caribbean that respond to natural

S. E. Galaitsi · C. Cummings · B. D. Trump (✉) · I. Linkov
Engineer Research and Development Center, US Army Corps of Engineers, Vicksburg, MS, USA
e-mail: benjamin.d.trump@usace.army.mil

S. E. Galaitsi
e-mail: stephanie.galaitsi@usace.army.mil

C. Cummings
e-mail: cummings@usace.army.mil

I. Linkov
e-mail: igor.linkov@usace.army.mil

C. Corbin
The Caribbean Environment Programme, United Nations Environment Programme, Kingston, Jamaica
e-mail: christopher.corbin@un.org

J. Keenan
School of Architecture, Tulane University, New Orleans, LA, USA
e-mail: jkeenan@tulane.edu

S.-A. Cox
Barbados Fisheries Division, Ministry of Environment and National Beautification, Green and Blue Economy, St. Michael, Barbados

G. Joseph
The Climate Resilience Agency for Dominica (CREAD), Roseau, Dominica
e-mail: gjoseph@creadominica.org

C. D. Metcalfe and E. R. Bennett (eds.), *Building Resilience to Climate Change in Small Island Developing States (SIDS) in the Caribbean*, Environmental Contamination Remediation and Management, https://doi.org/10.1007/978-3-031-37376-3_5

disasters have focused on reactive solutions that are limited in scope and under-resourced. This article reviews the opportunities to build resilience for Caribbean SIDS to not only "bounce back" from disruptions but also "bounce forward" to make long-lasting improvements to infrastructure and the economy, as well as the well-being of island citizens. This chapter critically examines current gaps and barriers to building resilience in Caribbean SIDS, including the need to develop a "resilience culture" among institutions and reduce reliance on external donors for funding. This chapter can assist policymakers in Caribbean SIDS to transition to programs for building resilience that deliver long-term benefits for the economy, social welfare, and environmental stewardship.

5.1 Introduction

Projections indicate that the Caribbean will experience disproportionately high risks from climate change (Thomas et al. 2020; Vosper et al. 2020) despite the minimal contribution to greehouse gase emissions. A variety of climate-related challenges and stressors are predicted for Caribbean SIDS due to their geographical position and densely populated coastal regions. Long-term stressors, such as sea level rise, ocean warming and acidification, and coastal erosion, as well as acute hazards, such as hurricanes, tropical storms, droughts, and flooding already affect these islands.

In addition, while hurricanes have accounted for about 95% of total natural disaster-related damage since the 1960s (Burgess et al. 2018), some Caribbean SIDS also face threats from seismic and volcanic activity. For example, the 2021 volcanic eruption in Saint Vincent and the Grenadines in the southern part of the Lesser Antilles displaced between 20,000 and 30,000 people (Andrews 2021; Global Volcanism Program 2021).

In addition to these acute hazards and disruptions, Caribbean SIDS face chronic impacts posed by increasing temperatures, sea level rise, increases in wave power and precipitation, among other environmental stressors (Thomas et al. 2020). This emphasizes the need for strategic approaches, projects, and actions to mitigate these risks. Consequently, scientific and policy discourse is increasingly focused on enhancing

P. McConney
Center for Resource Management and Environmental Studies, University of the West Indies, Cavehill Campus, St. Michael, Barbados
e-mail: patrick.mcconney@cavehill.uwi.edu

A. Cashman
AKWATIX Water Resource Management, St. John, Barbados

C. Springer
Caribbean Natural Resources Institute (CANARI), Sir Arthur Lewis Community College (SALCC), Castries, Saint Lucia
e-mail: springerc@comcast.net

system resilience to both acute and chronic stressors arising from evolving environmental and climatic conditions. However, few institutionalized and economically viable initiatives have been implemented to transform the growing knowledge and awareness of climate threats into resilience-based programs and interventions.

Several barriers hinder the successful execution of a multi-sectoral, multi-stakeholder resilience strategy, and create negative feedback loops that increase vulnerability to climate disruptions like extreme weather events (Linkov and Trump 2019). Some obstacles are scientific, such as the need for standardized language to define and parameterize climate resilience. Others emerge from challenges in translating resilience science to governments and societies, including an affordable and robust climate insurance and reinsurance industry. There is a need to cultivate and retain a skilled hazard management workforce, create and implement transparent policy mechanisms that monitor and adapt to shifting hazard cycles, and plan for investment in resilient infrastructure. While resilience research offers extensive prescriptive guidance to address various threats, there is a critical gap in integrating cutting-edge resilience knowledge with best practices and behavior change. Adhering to such practices could empower countries and regions to create robust environmental, land use, infrastructural, and economic policies and to effectively implement them through monitoring protocols and evaluation measures.

Caribbean SIDS, although diverse, share certain characteristics, including remote geographic locations, high coastline-to-land ratios, and the positioning of a significant portion of their populations and critical infrastructure near the coast (Vosper et al. 2020). Sea level rise and acute hazards like hurricanes, tropical storms, and flooding (Thomas et al. 2020) impact residents and critical infrastructure systems like major roads, air and seaports, utilities and telecommunication networks (Mycoo 2018; Vosper et al. 2020; Linkov et al. 2022a, b). For example, in September 2017, Hurricane Irma damaged or destroyed nearly 95% of structures on Barbuda and 90% on Saint Martin, and 90% of roads on Anguilla (Cangialosi et al. 2021). Two weeks later, Hurricane Maria, one of the deadliest hurricanes in recorded history struck the northeastern Caribbean, mainly impacting Dominica, Saint Croix, and Puerto Rico (Vosper et al. 2020). In September 2019, Hurricane Dorian rendered over 3,000 homes uninhabitable and damaged or destroyed hospitals, schools, and fisheries in the Bahamas (PAHO 2019).

The goal enshrined in the Paris Agreement of capping global warming at 1.5 °C still leaves Caribbean SIDS vulnerable to the growing severity of extreme weather events (Vosper et al. 2020). Under a 2 °C global warming scenario, projections indicate that the worldwide average intensity of Category 4 and 5 storms will increase by 5%, while the Caribbean will experience a 13% rise in intensity (Knutson et al. 2019; Spencer and Strobl 2020). The relationship between increased storm intensity and changes in severity or frequency remains unclear, but Caribbean SIDS are expected to face amplified climate hazards as temperatures and sea levels increase, even though their role in anthropogenic climate change is minimal (UNFCCC 2005).

Communities confronting hazards must also plan for long-lasting environmental, economic, and health repercussions, emphasizing the need for both immediate and extended recovery and adaptability. This is particularly vital for Caribbean SIDS, which require tailored climate resilience policies and practices that consider their unique social and economic circumstances. Key economic sectors such as agriculture, fisheries, and tourism rely heavily on the natural environment, but the economic stability of these industries is sensitive to environmental fluctuations (Thomas et al. 2020). By 2030, Caribbean SIDS are expected to experience economic risks higher than the global average of 0.5% for average losses in annual GDP, with projected losses ranging between 0.75 and 6.5% of GDP (Thomas et al. 2020; UN-OHRLLS 2015). This disproportionate impact on GDP stems from small population sizes, distance from global markets, elevated transportation costs, susceptibility to external economic shocks, and fragile ecosystems on land and at sea, which make SIDS particularly prone to impacts due to limited economic alternatives (UN-OHRLLS 2022). Additionally, the geographic isolation of Caribbean SIDS imposes high import and export costs that strain island economies during both stable and hazardous periods (UN-OHRLLS 2022).

Valuable insights can be gleaned from local and national climate resilience strategies that may be extended to other Caribbean SIDS. For example, after Hurricane Maria in 2017, Dominica's government developed a National Resilience Development Strategy and a Climate Resilience and Recovery Plan to improve the country's ability to withstand climate-related challenges. The plan aims to achieve sustainable communities, infrastructure, ecosystems, and economic growth while addressing present and future climate-related, climate-exacerbated, and non-climate challenges (Robinson and Butchart 2022). To achieve their explicit "Climate Resilience Targets" by 2030, Dominica has created 10 multi-jurisdictional, cross-sectoral initiatives that encourage behavioral change and intervention at all levels. These initiatives can be monitored for progress as they are quantifiable. The Climate Resilience and Recovery Plan is tailored to Dominica's unique socioeconomic and equity vulnerabilities, Indigenous (i.e., Carib First Nation) viewpoints, and public risk communication and perception insights. While Dominica's national climate resilience approach may be transferable to other Caribbean SIDS, it must be customized to meet the specific needs of different islands. In another example, the United Nations Environment Programme (UNEP) initiated the "Enhancing Coastal Ecosystem-based Adaptation in Small Island Developing States" project in Grenada. This project seeks to restructure, coordinate more effectively, and oversee Marine Protected Areas (MPAs) in Grenada to preserve ecosystems reliant on coral and mangrove health, while simultaneously benefiting and encouraging local stakeholders to participate in MPA ecosystem-based adaptation efforts (UNEP 2020). This UNEP initiative could potentially be expanded and applied to other SIDS coastal conservation zones.

This chapter examines the challenges facing Caribbean SIDS in their shift from vulnerability reduction to resilience building by analyzing current gaps and obstacles to resilience knowledge, requirements, and implementation. While limited research exists, several barriers to developing climate resilience and adaptation have been identified. For instance, a study by Robinson (2018), which involved interviews with senior Caribbean policymakers found that climate resilience policy and adaptive capacity were restricted by several factors, including lack of.

- financial resources,
- technical capacity,
- data availability,
- natural resources,
- human resources,
- knowledge of climate effects,
- national policy focus,
- monitoring and evaluation capacity,
- geographic size, and
- public education and awareness.

Land arability, social acceptability of proposed adaptations, and the frequency and intensity of climate impacts were the least commonly reported constraints (Robinson 2018). This chapter seeks to elaborate on these and other facilitators and barriers that impact the capacity of Caribbean SIDS to develop, adopt, and implement policies and practices for building resilience. Furthermore, this article emphasizes the unique requirements of small islands that may guide the strategic development of Caribbean climate resilience in the future.

Broadly, workshop participants envision the recovery and response process through approaches that empower SIDS and other communities to cultivate a culture of resilience, support policy adaptability, and deploy adequate tools and analytics to "bounce back" from disruptions while also prioritizing societal values like safety and wellbeing, equity and justice and sustainability, which can propel communities to "bounce forward" (Fig. 5.1). We consider these processes as opportunities that can be incorporated into traditional "recovery and response" phases of crisis planning and mitigation, as depicted in Fig. 5.1. The following sections of this chapter provide a comprehensive discussion of the ongoing needs for adopting climate resilience approaches in the Caribbean and identify opportunities and requirements for advancing climate resilience in this region.

5.2 Part 1: What Are the Barriers to Adopting Resilience in the Caribbean?

5.2.1 The Need to Foster a "Resilience Culture"

Because of the geologic and geographic vulnerability of Caribbean SIDS to natural disasters (e.g., hurricanes, flooding, volcanic eruptions), the residents of the region confront quasi-regular disruptions, and as such they are well acquainted with the need to develop resilience. Because systems cannot be designed to fully prevent disruptions

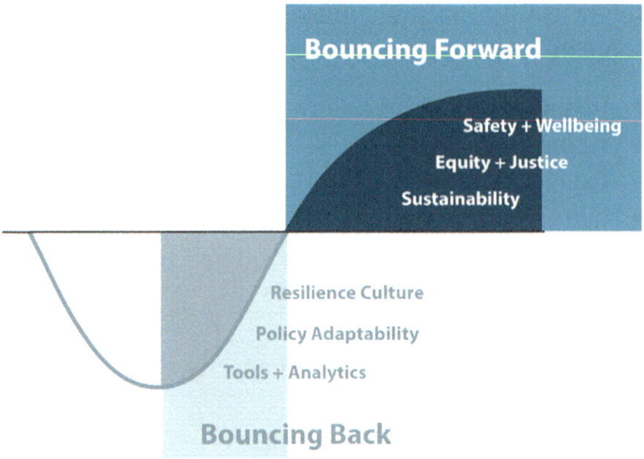

Fig. 5.1 Perspectives on building resilience in Caribbean SIDS by simultaneously "bouncing back" and "bouncing forward" from disruptions, with the specific competencies required to excel at both strategies

from occurring or creating impacts, stability over time requires the ability to recover. At the local scale, residents of all ages and in all situations encounter conditions that require recovery, perseverance, and adaptation. Their success generates communal pride in overcoming adversity stemming from these disruptions.

Despite successes in individual resilience, the system may not be well-served when the responsibility of recovery is left to individuals alone. Institutional resilience can reduce the burden on individuals by ensuring that services like electricity, solid waste collection, sewage treatment, and access to potable water are restored as quickly as possible. A country's failure to provide institutional resilience can place undue strain on individuals and local communities, and lead them to experience burn out and marginalization (Vercio et al. 2021). Efforts to support equity in the Caribbean have traditionally taken the form of low-cost housing, national health insurance, and improving employment opportunities. Expanding institutional planning to include building resilience to disasters can also support equity.

Island institutions can leverage an individual and local "resilience culture" to help establish resilience planning on a societal scale. However, even when resilience planning measures are developed in the Caribbean, their implementation often stalls since the public institutions often do not possess the tools and resources to enact these plans. Moreover, the frequency and magnitude of disruptions hinder the capacity to formulate proactive rather than reactive measures. This reality partially results from political emphasis on more immediate problems in Caribbean SIDS like unemployment or the challenges of developing tourism (Holladay and Powell 2013). While actions like protecting a mangrove forest can improve resilience to flooding and coastal damage, the resources needed to stabilize such ecosystems may be hard to justify when compared to nearer-term benefits like resort construction. Many governments in Caribbean SIDS make their investment plans in 3–5-year increments, and some large-scale or longer-term resilience projects may not have enough time to mature and provide returns on investment to the government that implemented them. Where resilience to uncertain future disasters requires forgoing certain profit, the short-term benefits are likely to prevail. This curtails efforts to pursue long-term resilience planning. Furthermore, what resilience culture that exists at a broader political or governance level is often not strong enough to counteract the allure of short-term gains, even when these projects jeopardize long-term benefits. Workshop participants noted that this pattern has existed for decades in Caribbean SIDS planning. Solutions that were viewed favorably decades ago remain unrealized today, long after their benefits should have been felt.

Gaps in resilience implementation also arise from the interconnectivity of Caribbean island economies at regional and sub-regional levels. Certain economic sectors like tourism and agriculture dominate Caribbean SIDS, and this translates to a requirement for government projects to have commitments from the private sector. Many consultants' recommendations, actions, and policies can remain unimplemented, partially due to insufficient participation from the private sector and/or a lack of public support (Scobie 2018). It is imperative that the private sector better engage with the process of policymaking because businesses are not insulated from long-term impacts. Governments can also strive to demonstrate that resilience planning and culture will positively affect profits for the private sector over time as well as offer improvements to broader society. The private sector tends to respond to economic incentives, but these do not always align with resilience planning because of a focus on shorter timeframes and less interest in mitigating against events that are unlikely in the short term. To implement resilience planning, Caribbean SIDS will need to foster an appreciation of resilience across all sectors.

Caribbean SIDS often receive support from international donors to recover after disasters. However, donors can impose their own prerogatives on infrastructure projects, such as a focus on building or improving bridges, harbors or electrical grids, and this may constrain the ability of Caribbean governments to invest in projects that support resilience and perhaps have higher upfront capital costs. As described in other chapters in this book, these donors may also lean towards supporting engineered or "grey" infrastructure projects and not nature-based solutions for building resilience. The governments of Caribbean SIDS will need to convince donors like the World Bank and the Inter-American Development Bank that investments in building resilience should be a priority.

Despite the challenges in implementing resilience projects in the Caribbean SIDS, planners from this region have many advantages over planners elsewhere in the world because of their broad experience and understanding of resilience. Additionally, the frequent occurrence of disasters around the Caribbean has armed residents with personal experience in building resilience. These experiences also viscerally demonstrate the limits of risk management. The massive scale of the disasters precludes "hardening" that could fully insulate inhabitants from the impacts (Linkov et al. 2018). Planning for resilience requires a departure from the mindset of robustness/risk management, but it is one that Caribbean Island residents are likely to better understand and value than professional planners elsewhere in the world. Elements of a resilience culture exist in the Caribbean, but it needs to be extended to institutions to internalize and codify it and to monitor and evaluate its effectiveness over time.

Returning to Dominica's example, the value of building resilience is already enshrined within Dominica's planning policies. Even though rebuilding to climate-resilient standards costs 25% more (Dorst 2021), the benefits have already been demonstrated by the construction of infrastructure that is able to resume service delivery faster after storms (e.g., communications networks). There is some intermingling of resilience and risk management concepts in Dominica's framework. Buildings that stand up to Category 5 hurricanes are robust (sufficiently hardened) but are not necessarily resilient. For example, building bridges higher to avoid debris overflow is similarly not resilience, but risk avoidance. However, the institutional investments being made in health, education, and food security may enhance Dominica's resilience over time, though the total effects will take some years to be manifested measurably. Dominica's policy focus on resilience represents an example of the translation of resilience culture into institutional implementation in the Caribbean islands.

5.2.2 The Need for Institutions with Practical Tools to Craft and Implement Adaptable Resilience Policies

Caribbean SIDS residents must repeatedly respond to complex hazards that affect their infrastructure, economy, ecosystems, and communities. Yet, these responses are typically reactive and ad hoc. There are ample opportunities to develop and implement more proactive and adaptable climate resilience policies. Positioning small islands to strategically plan for and adapt more quickly to diverse climate-related threats requires a robust climate resilience culture and corresponding policy initiatives. However, as described by Palma-Oliveira et al. (2018), the barriers to institutional climate resilience, include (a) an emphasis on short-term policy solutions, (b) policy development not translating to actionable implementation, (c) inconsistent incentives that fail to unify public, private and community stakeholders towards adopting climate resilience policies and practices, and (d) a dearth of data and information and an inability to apply data and information to guide policy and to evaluate its effectiveness over time.

Diverse natural hazards threaten the economies, infrastructure, and public health and safety within Caribbean SIDS. The costs of natural hazard response and recovery are roughly 7–19% of annual GDP for Caribbean SIDS (Galindo et al. 2022). Such high costs of climate-related response and recovery efforts may cause policymakers to adopt only least cost solutions, with only short-term benefits. Emphasizing short-term payoffs, however, poses the risk of shortcuts that become vulnerabilities in the face of repeated future damages and recurring recovery costs. Instead, policies can embrace climate resilience by strategically planning for and adapting to the changing climate. Policy development should use longer time horizons to consider and weigh the costs and benefits of policy and infrastructure alternatives under a range of potential stress scenarios (Linkov et al. 2022a, b). This can help identify policy alternatives that may be more adaptable, robust, and resilient to future disruptions, while requiring lower maintenance costs during their life cycle. Such policy alternatives may impose higher initial capital costs, but the investments should be more cost-effective over time.

Further, it may be necessary to consider factors beyond direct monetary costs in developing climate resilience policy. Improved resilience can allow critical services to recover quickly after a storm; this affects community wellbeing and can also save lives (Peterson 2020). Using longer timeframes to compare cost-effectiveness and infrastructural integrity may also demonstrate the benefits of climate-resilient investments to donors. Communicating long-term benefits to internal and external stakeholders will be vital for developing the resilience policies needed for both strategic planning and climate adaptation. Such an endeavor requires a well-established framework for monitoring and for communicating results to multiple stakeholders.

Supporting climate resilience will require developing and implementing strategic policies. Currently, as in much of the rest of the world, policy development in the Caribbean islands can be an extended process vulnerable to bureaucratic setbacks. Moreover, whatever new and amended policies that emerge may not be implemented consistently, if at all. Resilience policies must be balanced with practical, on the ground solutions. Climate resilience efforts often exist in silos; for instance, coastal restoration projects may be implemented when a more holistic "ridge to reef" approach is warranted. Decision makers can focus on identifying objectives and establishing actions at national, regional, and local levels. Behavioral shifts across various stakeholders may lose momentum, in part because indicators related to progress in achieving those goals are not transparent. Further, the policies that do exist are not consistently enforced or regulated, particularly for the private sector, and this contributes to the gap between policy development and policy implementation. Climate resilience education can support both regulating and incentivizing climate resilience policies. This aspect will be discussed in the next section.

Finally, climate resilience policies must be developed and implemented in ways that are scalable to multiple sectors and jurisdictions. Project implementation will benefit from harmonizing multiple decision-making scales, including sub-regional and regional scales. Although national governments may be best suited to support these types of projects, collaboration is needed from other scales, including at the individual- and household-levels. In addition to long-term planning, adaptation and development of actionable policy goals, climate resilient culture and effective climate resilience policies rely on the behavioral choices of individuals. Public support of projects could be galvanized by publicity campaigns emphasizing policy benefits, even when the benefits are uncertain or more likely to arise over the very long term. Such benefits should be clearly communicated, especially in cases where people perceive projects as representing an imposition, such as changing their behavior or if there is a trade-off that negates a more immediate tangible benefit.

In summary, developing and implementing climate resilience policy will require institutions to consider: (a) long-term strategic planning, (b) transitions from policy development to actionable implementation, (c) multi-stakeholder engagement, including with community members, and (d) institutionalized monitoring, evaluation, and feedback regarding policies and actions.

5.2.3 The Need for Evidence-Based Decision-Making to Guide Response and Recovery to Disruptions

Fostering a resilience culture and invigorating resilience policy will lay a strong foundation to maximize economic and governance opportunities (Hynes et al. 2022). However, the success of these key priorities relies on the quality of information and knowledge-based data. The days of governance through expert intuition are gone. Today the best governance and business leadership and practices are informed through robust evidence-based decision-making that guides resilience planning, response, and recovery from disruptions (Contreras and Gloria 2017).

5.2.4 Sustaining Knowledgeable Workforces and Collaborations

Evidence-based decision-making is ultimately a human-centered activity that requires a skilled and adaptable workforce capable of filling the booming demand for reliable and valid information about future conditions (Pagán-Trinidad et al. 2019). Planning, responding, and recovering from natural disasters involves multiple fields of study, including climatology, geological and environmental science, civil and environmental engineering, economics, risk analysis and management, public health, and risk communication. Meeting the resilience data demands of tomorrow requires investment today in explicit interdisciplinary education and training programs.

Independent nations and territories should identify opportunities for strategic education partnerships to train the workforce of tomorrow for the interdisciplinary activities required to support the acquisition of data and information. Private and public investment in state-of-the-science curricula across all education levels (kindergarten through post-secondary levels) can invigorate teaching and learning about climate issues and infuse authentic inquiry of pressing societal issues into existing education rubrics. Nonetheless, obstacles exist such as limited funding for curriculum development and equipment, insufficient technical proficiency and expertise among educators, and insufficient time to create and deliver interdisciplinary lesson plans (World Bank 2021). Such barriers affect the ability of educators to adequately teach interdisciplinary science and entrench student disinterest in pursuing training in science and technology as they mature through the educational system.

To cultivate a skilled and motivated workforce capable of addressing major challenges, innovative educational programs may be required to break this cycle. Similar groundbreaking educational initiatives are already being implemented in other fields, such as applied biotechnology education, where mobile science laboratories and customized interdisciplinary curricula are being utilized. Such laboratories can reach diverse student populations more fairly across regions, fostering growth and innovation while preparing the next generation of the workforce. Another chapter in this book describes some of these educational activities in the US Virgin Islands.

5.2.5 Data Needs, Comparative Forecasting Tools, and Real-Time Data Analytics

Effective climate resilience management warrants robust data collection about current statuses and needs, as well as the capacity to synthesize data into knowledge and insight that will inform decisions. However, utilizing data during a crisis or disaster can be especially challenging as circumstances are unclear and change rapidly. Emergency responders require numerous alternatives for employing multi-level platforms to gather data and evaluate on-site situations, even in real-time as events occur (Maconi 2021). Data collection technologies have the potential to significantly enhance emergency response and recovery efficacy. Data infrastructure and research platforms can be customized to answer three fundamental questions for climate resilience planners (Becker et al. 2021):

- What do we know, based on prior events? Historical data can help identify vulnerable areas and the needs for parameters and data that will guide future decisions.
- What climate-based disruptions are expected? Prior to mobilizing climate resilience plans, desktop analyses may be used to proactively assemble local and historical data to identify communities or geographies that have particular vulnerabilities.
- What changes on the ground can improve data collection? Additional infrastructure to enable rapid data collection may be reactively deployed at large-, mid-, and small-scales. These include remote sensing applications, Unmanned Aircraft Systems (UAS), and mobile phone apps or other field collection devices.

These questions can form the foundation for creating a set of instruments and tools to quickly gather data at suitable levels to advise decision-making. Such data can better support disaster response professionals by reducing the need for labor and travel, and by helping to overcome common logistical challenges in disaster zones (e.g., road blockages, power outages, absent homeowners). Web-based tools can help emergency managers to rapidly access community data and summary metrics during unfolding crises and disasters. On-the-ground tools, such as smartphone applications, are essential for collecting field data that are not observable at larger scales. These tools can validate information collected via remote sensing and can address data gaps. Integrating these tools into data management structures and workflows is crucial to align and synthesize nested data sets across different scales. A well-designed data structure can provide a common operating framework that enables easy visualization, analysis, and coordination. Such a structure can support on-the-ground missions and facilitate after-the-fact study, simulation, and comparative forecasting of future issues.

5.2.6 Managing Cascading Climate Resilience Risks

Traditional risk assessment and management focuses on individual threats using historical data or predictive models, while all-hazard frameworks consider multiple threats. However, this approach overlooks the interconnections of real systems, such as urban environments and climate disruption, and the non-linear and cascading nature of some threats, such as hurricanes (Romero-Duque et al. 2020). To address this, future data usage should integrate real-time decision-making and synthesis of trade-offs in policy objectives, accounting for the possibility of cascading threats and climate disruptions. It is necessary to develop new, adaptable, and dynamic evaluation tools that can adjust inputs and outputs for each case. Decision-making frameworks based on evidence should prioritize the impact of cascading and non-linear climate disruptions on affected populations and infrastructure stability.

Together, these foundational concepts of cultivating a resilience culture, prioritizing policy adaptability, and developing adequate tools and analytics function together to enable communities to "bounce back" from disruption (Asariotis 2020). We envision these concepts as the primary "pieces of the puzzle" that will build climate resilience by design and help to foster systems that enable SIDS to "bounce forward" amidst climate disruptions, as depicted in Fig. 5.2.

5.3 Part 2: What Are the Opportunities and Necessities for Resilience Moving Forward?

Effective policy for climate resilience is challenged by common problems across Caribbean nations, such as (a) the need for a holistic conceptualization of the various infrastructural, environmental, and societal systems affected by climate stressors, (b) the challenge of shaping this conceptualization into actionable policy, (c) implementing those policies in a transparent manner, with expectations for monitoring and improvement, and (d) applying climate policies to various levels of scale, from households to consortia among multiple nations (e.g., CARICOM countries, US Territories). Increasingly, there is agreement that building resilience, and particularly fostering systemic capacity for recovery and adaptation post-disaster, are policy necessities. Yet, how they might be operationalized remains unclear.

While such policies might take many forms, they confer resilience through two general avenues: by-intervention (i.e., resilience by intervention), and by-design (i.e.,

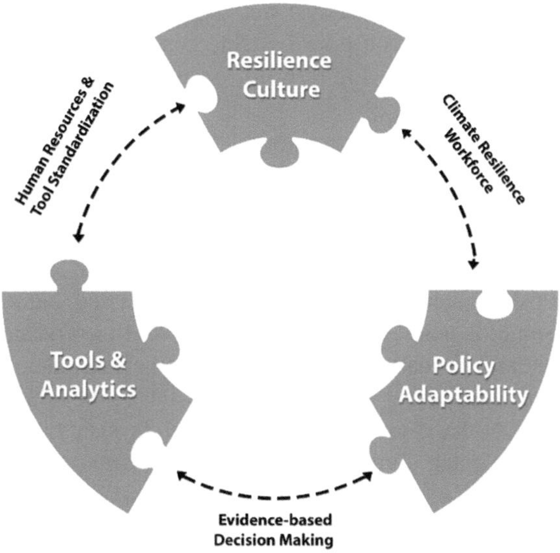

Fig. 5.2 "Pieces of the puzzle" that underlie resilience by design (RbD) and the resources required to foster systems that will enable Caribbean SIDS to "bounce forward" from disruptions

resilience by design). Resilience by Intervention (RbI) approaches rely upon transfers of energy and resources from one body to another (e.g., from other countries, philanthropic organizations or donors, multi-national emergency response agencies, etc.). Historically, interventionist approaches have been used to overcome substantial disruptions to infrastructure and economic systems post-disaster but are often time-limited and do not address the long-term implications of recovery (Berke and Beatley 1999). Likewise, interventionist approaches are often reactive, which leaves the islands equally vulnerable to subsequent disasters and additional reconstruction costs (Galindo et al. 2022). The recurring nature of climate risk to Caribbean SIDS dictates that even the most successful resilience culture will sometimes benefit from the additional resources. Some of these resources could be directed to assure that the islands have more built-in resilience capabilities through "resilience by design" (RbD).

RbD refers to the capacity of communities and nations to absorb, recover from, and adapt to climate shocks by the grace of their own knowledge, capabilities, and resources. By-design resilience is fundamentally scalable, requiring recovery-driven philosophies from individual to national/multi-island perspectives. These approaches position the economies of Caribbean islands and their societies to stimulate recovery quickly, even before external support can be operationalized. By-design climate resilience supports the efficient recovery of local economies, local finance, consumer spending, etc. and can prevent a cascade of problems that affect other systems, such as utility operations, health services, overall societal health, and wellness.

Achieving both RbI and RbD requires incorporating principles of resilience into various levels of society, as well as defining what systems must be resilient to disruption and what types of outcomes are desirable post-disaster. Common endpoints include: (i) the systemic capabilities needed to restore basic service delivery and survival, and (ii) the systemic capabilities needed to facilitate recovery scenarios that improve upon prior baseline conditions. These capabilities denote "bouncing back" and "bouncing forward," respectively, and we explore both concepts further below.

5.4 "Bouncing Back"

Post-disaster, an acute threat may have passed, but it leaves destruction in its wake. Some of this destruction may be poised to produce further harm as people cope without critical services like electricity, clean water, sanitation, solid waste disposal, and safe housing. Residents that weathered the initial disaster can subsequently face heat stroke, disease transmission, and physical hazards, among many others threats to wellbeing. Furthermore, the same system outages affecting residents also affect important infrastructure, like hospitals, schools, communication networks, and transportation corridors serving the supply chain (Mohammed et al. 2023). For instance, the impacts of Hurricane Maria on already-vulnerable systems in Puerto Rico, particularly the power outage that reduced hospital services, telephone communications, and government response capabilities, were attributed as the cause of over 3,000 deaths. Hospital outages in the Bahamas after Hurricane Dorian accounted for at least one death.

When critical systems fail, the effects can cascade into other systems that would not have otherwise been affected. For example, a person might become ill from consuming unsanitary water, but this can become life-threatening if that person cannot access healthcare at the local hospital due to power outages or lack of medicine delivery. The resilience of "bouncing back" implies getting critical systems functioning at a sufficient level to prevent such cascades (Sandoval and Sarmiento 2020). Thus, a society with capacity to bounce back is resilient in the face of acute disruptions, with known resources ready to be distributed in efficient ways to ensure rapid recovery after an event. Bouncing back refers to a recovery in which critical systems come back online quickly to address the emergency needs of local population and systems.

To facilitate bouncing back after a disaster, priorities should include ensuring a continuous supply of emergency items such as food and medical supplies in the supply chain and ensuring that hospitals have backup generators with sufficient fuel. In areas where bridges are likely to be affected, alternative routes should be mapped, such as backroads that can accommodate smaller trucks. In addition, smaller trucks should be equipped to navigate such backroads to transport emergency items that are smaller in volume than normal shipping capacity. Moreover, accommodations and necessary infrastructure should enable emergency transport via small cargo planes. These measures will ensure that critical services and resources can be provided even while a broader and more permanent recovery process is underway.

Climate finance plays a critical role in supporting Caribbean SIDS as they implement resilience strategies to adapt to the impacts of climate change. Climate finance refers to the funding directed toward climate change mitigation and adaptation initiatives, which can come from various sources, including public, private, domestic, and international entities. For Caribbean SIDS, access to climate finance is essential for developing and implementing robust climate resilience strategies because the island nations often lack the resources and technical capacity to address climate-related challenges on their own. By leveraging climate finance, Caribbean SIDS can invest in innovative solutions and technologies that build resilience against climate threats, while promoting sustainable development and enhancing the overall wellbeing of their populations (Brichetti et al. 2021).

5.4.1 Climate Finance

One strategy that contributes to the overall narrative of SIDS and climate resilience on climate is the development of innovative financial mechanisms for mobilizing resources (Keenan et al. 2021). For instance, blue bonds and green bonds are innovative financing tools that can help raise funds for climate resilience projects in Caribbean SIDS. These bonds are issued by governments or corporations to finance environmentally friendly projects and can attract investment from both public and private sectors, thereby increasing the pool of available resources for climate resilience initiatives. By tapping into these innovative financial mechanisms, SIDS can access the necessary funding to invest in cutting-edge technologies and infrastructure projects that enhance climate resilience, such as renewable energy systems, coastal protection measures, and sustainable water management practices.

Another idea that could support climate finance for Caribbean SIDS is the implementation of results-based financing (RBF) mechanisms for climate resilience projects. RBF mechanisms are designed to link financial incentives to the achievement of specific, measurable outcomes related to climate resilience, such as reduced greenhouse gas emissions, increased ecosystem resilience, or improved community preparedness (Steckel et al. 2017). By using RBF mechanisms, SIDS can ensure that climate finance is used effectively and efficiently to achieve tangible results, while also promoting transparency, accountability, and evidence-based decision-making in climate resilience initiatives. Furthermore, RBF mechanisms can help attract private sector investment by offering a clear value proposition and reducing perceived risks associated with resilience projects. By integrating RBF mechanisms into their resilience strategies, Caribbean SIDS can optimize the use of climate finance to achieve their adaptation and mitigation goals, while also fostering innovation, collaboration, and stakeholder engagement in the pursuit of a more sustainable and resilient future.

5.5 "Bouncing Forward"

After bouncing back meets the needs for survival post-disruption, a significant question remains: is the current systemic configuration one that meets the needs of broader society? If not, what normative goals should nations experiencing chronic disruptions strive to achieve as they progress through post-disaster recovery and adaptation?

The recovery and adaptation process is rife with opportunity to rebuild physical assets (infrastructure, urban environment) and reconfigure incentive structures (economy, society). Rather than recover to an initial baseline, many cities and countries affected by disasters have instead "bounced forward" or invested resources allocated for the recovery process to achieve normatively preferable goals (Muñoz-Erickson et al. 2021). Examples include significant investment into jobs diversification and training for households in Japan affected by the Fukushima Nuclear Disaster (Zhang et al. 2019), or reconfiguring policies to anticipate climate stressors to Caribbean nations post-Hurricanes Irma and Maria (Moulton and Machado 2019).

Post-disaster goals can vary significantly, based upon the capabilities and resource limitations of a society under disruption. Many normative post-disaster goals are relatively modest and may focus on how infrastructure or local environments can be managed and utilized in new ways by society. However, more transformative post-disaster goals can seek to change the types of work or financial incentives that a local area can sustainably support over longer time frames. Thus, bouncing forward emphasizes the priorities in resilient responses that best position assets, communities, or other features to be better prepared for the future (Wentworth et al. 2020). This involves both adapting to the circumstances that are projected to be permanently altered over time due to climate change or other social effects, as well as being resilient to acute disruptions (Trump et al. 2017). Examples could include moving electrical wires underground to reduce vulnerability from wind, or re-populating natural or agricultural environments with more resilient or native vegetation.

Furthermore, workshop participants emphasized that disaster responses should incorporate broader societal needs and values, including addressing equity, wellbeing, and sustainability. Below we explore each in turn and identify how RbD can be achieved through synergistic governance activities that prioritize societal needs.

5.5.1 *Safety and Wellbeing*

The fundamental duty of a government is to safeguard its citizens. Disasters can severely disrupt societies across various sectors, such as the economy, politics, education, health, innovation, arts, science and technology, and more. Therefore, when allocating resources, decision-makers must balance competing needs by prioritizing short-term responses alongside long-term societal planning. During times of disruption, it is crucial to prioritize the physical, psychological, and social wellbeing of the broader society to ensure societal functioning. Protecting personal safety is of utmost importance, and resilience planning should include emergency life-saving measures such as anticipatory evacuations, enforced curfews and lockdowns, and potentially relocating high-risk populations. Ensuring the safety and contentment of communities can lead to greater satisfaction with their lives, which may contribute to societal stability and sustainability.

To promote safety and wellbeing within a resilience strategy, there is a need for early warning systems (EWS) for climate-related hazards. EWS are essential tools for reducing the impacts of extreme weather events on communities because they provide timely and accurate information about impending hazards, thus enabling individuals and communities to take necessary precautions and actions to protect themselves (Bonato et al. 2022). By investing in and implementing EWS, Caribbean SIDS can improve the safety and wellbeing of their populations by enhancing preparedness, reducing response times, and minimizing the loss of life and property during disasters. Furthermore, incorporating advances in remote sensing, data analysis, and risk communication can help make EWS more effective and accessible for diverse communities, ensuring that everyone has access to critical safety information.

Nature-based approaches for passive protection against climate risk that could promote safety against a range of possible environmental and climate shocks include mangrove restoration, coral reef conservation, and coastal reforestation. These approaches can provide significant benefits for safety and wellbeing by reducing the impacts of climate-related hazards, such as storm surges, coastal erosion, and flooding. These natural systems not only serve as buffers against climate threats but also provide essential ecosystem services that contribute to the health and wellbeing of communities, such as clean air, water filtration, and recreational opportunities (Mercer et al. 2012).

5.5.2 *Equity and Justice*

After a disaster, decision-makers must prioritize responses for the recovery of many different sectors of society. While a traditional cost–benefit analysis in Caribbean SIDS might emphasize the importance of hotels and the tourism industry, an equitable response takes a more holistic view that considers factors like historical inequities, differences in resource access, vulnerable and disadvantaged members of society, as well as need. Prioritizing responses and recovery using an equity lens will address pre-existing imbalances and ensure that all end points after the recovery are on more equal footing than before the disaster occurred (Cortés et al. 2020). Ensuring equitable responses can be challenging, as it requires planners to identify barriers, some of which may be invisible, and some of which may have been intentional due to a legacy of prioritizing other goals. The existence of inequity may also reflect pre-existing biases of society or the legacy of colonialism, and thus striving to do better may not be well understood or valued by a broader audience. However, equity in disaster responses not only provides resilience for the communities that most need it, but also could build better capacity for RbD in the future (Sheller 2021).

Actionable strategies to address equity and justice challenges in Caribbean SIDS include the development and use of vulnerability assessments that help identify and prioritize at-risk populations and ecosystems within a community, considering factors such as socioeconomic status, access to resources, and exposure to hazards (e.g., hurricanes, volcanic eruptions, tsunamis) (Fuchs et al. 2012). By incorporating vulnerability assessments into resilience planning and decision-making processes, Caribbean SIDS can ensure that their resilience strategies are inclusive and equitable and that they address the needs of the most vulnerable groups and areas. This approach promotes a more just and fair allocation of resources and support, thus fostering stronger, more resilient communities.

Another possibility includes the integration of community-based adaptation (CBA) initiatives. CBA is an approach to climate change adaptation that prioritizes local knowledge, participation, and empowerment, recognizing that communities have valuable insights into their specific vulnerabilities and capacities (Shammin et al. 2022). By actively involving local communities in the design, implementation, and evaluation of climate resilience strategies, SIDS can ensure that these efforts are culturally appropriate, sensitive to local needs, and equitable in their distribution of benefits. CBA initiatives can also facilitate capacity building and knowledge exchange among community members, empowering them to take ownership of their resilience and better prepare for future climate-related challenges.

5.5.3 Sustainability

When implementing RbD through allocation of resources for building resilience, decision-makers can strive to create a system that is self-sustaining. This would involve providing resources in such a way that RbD removes the need for RbI; that is, the resources provided are sufficiently durable or self-replicating that the receiving community will have more capacity to recover from subsequent disruptions without the need for an influx of external resources and funding.

One concept that contributes to the overall narrative of SIDS and climate resilience strategy is the implementation of "Nature-based Solutions" (NbS). As outlined in other chapters in this book, NbS can harness the power of natural ecosystems, such as healthy coral reefs, mangroves, and seagrass beds to mitigate the impacts of climate change, protect coastal communities, and support biodiversity. By investing in the restoration, protection, and management of these ecosystems, Caribbean SIDS can develop self-sustaining resilience measures that not only offer long-term benefits but also enhance the resilience of the receiving communities. NbS can help reduce reliance on external resources while contributing to the overall health and wellbeing of both people and the environment (Oen 2019).

Another option to promote sustainable and resilient island infrastructure and economies for Caribbean SIDS is the integration of renewable energy technologies and energy efficiency measures (Scobie 2018). Climate change not only exacerbates environmental hazards but also impacts the energy sector, thereby threatening the stability of energy systems and the economic growth of island nations. By transitioning to locally available renewable energy sources such as solar, wind, and wave energy, Caribbean SIDS can reduce their dependence on imported fossil fuels, improve their energy security, and create more resilient energy systems. Additionally, implementing energy efficiency measures can help reduce energy consumption and costs, further enhancing the self-sustaining capacity of these island communities.

Lastly, the adoption of climate-smart agriculture (CSA) practices can greatly contribute to climate resilience in Caribbean SIDS. CSA encompasses a set of agricultural practices that sustainably increase productivity, enhance adaptation to climate change, and reduce greenhouse gas emissions (FAO 2021). These practices include agroforestry, crop diversification, and water conservation techniques. By adopting CSA, Caribbean SIDS can improve their food security, reduce reliance on imported food, and enhance the adaptive capacity of their agricultural sectors. In turn, this will contribute to the development of self-sustaining resilience measures, ensuring that communities can better withstand and recover from future climate-related disruptions. These are but a few of many possibilities that can lead to island sustainability, safeguard against climate risk and promote new opportunities for local citizens in the coming years.

Leadership best practices can incorporate societal issues along with foundational governance needs to foster more robust and resilient systems. Prioritizing these broader societal needs as valuable "pieces of the puzzle," as illustrated in Fig. 5.3, can further enable SIDS to achieve resilience and should be considered early and often throughout the planning process.

Fig. 5.3 Integrating the "pieces of the puzzle" to achieve resilience by design (RbD)

5.6 Conclusions

Caribbean SIDS face unique challenges due to climate change and other environmental hazards. Addressing these challenges requires a paradigm shift from a vulnerability reduction mindset to one focused on building resilience, considering both local and regional perspectives. By learning from existing climate resilience policies and practices, and examining the facilitators and barriers to their adoption, Caribbean SIDS can develop tailored strategies that address their specific vulnerabilities, while leveraging their unique strengths and resources (Biosca Amat et al. 2022).

A key takeaway from this article is the importance of a multiscale, cross-sectoral approach to build climate resilience in Caribbean SIDS. By integrating local knowledge, Indigenous perspectives, and public risk communication, these island nations can develop more effective and context-sensitive strategies (Charles et al. 2022). Furthermore, fostering a culture of resilience and policy adaptability, with an emphasis on safety, wellbeing, equity, justice, and sustainability, will enable Caribbean SIDS to not only bounce back from climate-related disruptions but also to bounce forward, seizing opportunities for growth and development in the face of adversity.

As Caribbean SIDS continue to experience the impacts of climate change, it is crucial to engage in collaborative efforts among all stakeholders, including governments, the private sector, local communities, researchers, NGOs, and international organizations. By sharing experiences, best practices, and lessons learned from climate resilience initiatives, these stakeholders can support one another in developing and implementing effective adaptation strategies. Fostering an environment of innovation, experimentation, and learning will be essential for driving transformative change and building resilience in the Caribbean region. Equally important, however, is the need to promote cost-effective organizational and labor conditions, and building institutions and protocols that mitigate damage, promote expeditious recovery, and make SIDS more attractive for economic development. Sustainable finance is a key need to ensure the long-term growth and development of societal wellbeing in Caribbean SIDS as they are impacted by future climate stressors.

Caribbean SIDS have the potential to become leading examples of climate resilience and adaptation, showcasing the power of collective action and community-driven solutions. By proactively addressing the challenges of climate change, these island nations can protect their unique natural environments, cultural heritage, and socioeconomic wellbeing while advancing sustainable development for current and future generations (Vergara et al. 2013). The journey towards climate resilience in Caribbean SIDS is not only an urgent necessity but also an opportunity to build societies that can be support their members in a changing future.

Acknowledgements We thank the Centre national de la recherche scientifique based in Paris, France, the US Army Corps of Engineers, and the National Oceanic and Atmospheric Administration of the USA for funding to support the workshop held in Barbados in June 2022 that was the basis for the preparation of this chapter. The authors thank Donnell Cain and Ricardo Marshall for their contributions to workshop discussions, which informed the content of this paper. The opinions expressed herein are solely those of the individual and do not necessarily reflect the views or opinions of any organization, institution, or entity mentioned or associated with the individual. The statements made are personal viewpoints and should not be attributed to any organization with which the individual may be affiliated. Any references to organizations are for informational purposes only and should not be interpreted as endorsements or official positions of those organizations. The individual takes full responsibility for the opinions expressed, and readers or listeners are encouraged to form their own judgments and interpretations.

References

Andrews RG (2021) Why the volcano erupting in the Caribbean has such a deadly reputation. National Geographic, Washington, D.C., USA. Available on-line from: nationalgeographic.com

Asariotis R (2020) Climate change impacts and adaptation for coastal transportation infrastructure: a sustainable development challenge for SIDS in the Caribbean and Beyond. In: Wang Y (ed) Coastal and marine environments. CRC Press, Boca Raton, FL, USA, pp 253–264

Baptiste A, Martyr-Koller R, Pringle P, Rhiney K, Thomas A (2020) Climate change and small island developing states. Annu Rev Environ Resour 45:1–6

Becker A, Hallisey N, Bove G (2021) Toward regional hazard risk assessment: a method to geospatially inventory critical coastal infrastructure applied to the Caribbean. J Infrast Preserv Resil 2:13

Berke PR, Beatley T (1999) After the hurricane: linking recovery to sustainable development in the Caribbean. Disaster Prev Manag 8:370–452

Biosca Amat N, Rojas Bonilla L, Parkinson J (2022) Using climate risk and vulnerability assessments to prioritise Caribbean road investments. In: Proceedings of the institution of civil engineers—engineering sustainability 175, No 5, Thomas Telford Ltd, London, U.K., pp 224–235

Bonato M, Sambo B, Sperotto A, Lambert JH, Linkov I, Critto A, Torresan S, Marcomini A (2022) Prioritization of resilience initiatives for climate-related disasters in the Metropolitan City of Venice. Risk Anal 42(5):931–952

Brichetti JP, Mastronardi L, Rivas ME, Serebrisky T, Solís B (2021) The infrastructure gap in Latin America and the Caribbean: investment needed through 2030 to meet the sustainable development goals. Latin Am Policy J 12. https://lapj.hkspublications.org/the-infrastructure-gap-in-latin-america-and-the-caribbean-to-reach-the-sustainable-development-goals-by-2030/

Burgess CP, Jones J, Spencer N, Stephenson TS, Taylor MA (2018) Estimating damages from climate-related natural disasters for the Caribbean at 1.5° C and 2° C global warming above preindustrial levels. Reg Environ Change 18:2297–2312

Cangialosi JP, Berg R, Latto AS (2021) National Hurricane Center tropical cyclone report: Hurricane Irma (30 August–12 September 2017). https://www.nhc.noaa.gov/data/tcr/AL112017_Irma.pdf

Charles SH, Chang-Richards A, Yiu TW (2022) Providing a framework for post-disaster resilience factors in buildings and infrastructure from end-users' perspectives: case study in Caribbean island states. Int J Disaster Resil Built Environ. https://www.emerald.com/insight/1759-5908.htm

Contreras C, Gloria T (2017) An envision rating system approach to sustainable infrastructure in Latin America and the Caribbean. In: Proceedings of the international conference on sustainable infrastructure, pp 20–31. https://doi.org/10.1061/9780784481196.003

Cortés LM, Rodriguez-Rivera MZ, James JJ, Cordero JF (2020) The Caribbean strong summit: building resilience with equity. Disaster Med Public Health Prep 14:155–157

Dorst S (2021) Dominica develops resilience. International Monetary Fund. https://www.imf.org/en/News/Articles/2021/10/27/na110221-dominica-develops-resilience

FAO (2021) Climate smart agriculture: case studies 2021, Projects from around the world. Food and Agriculture Organization of the United Nations, Rome, Italy. https://www.fao.org/policy-support/tools-and-publications/resources-details/en/c/1469956/

Fuchs S, Birkmann J, Glade T (2012) Vulnerability assessment in natural hazard and risk analysis: current approaches and future challenges. Nat Hazards 64:1969–1975

Galaitsi S, Corbin C, Keenan J, Cox S, Joseph G, McConney P, Cashman A, Springer C, Cummings C, Trump B, Linkov I Balancing climate resilience and adaptation for Caribbean SIDS: promoting institutional capacities. Submitted to IEAM

Galindo LM., Hoffman B, Vogt-Schilb A (2022) How much will It cost to achieve the climate goals in Latin America and the Caribbean? IDB Working Paper Series Nº IDB-WP-01310 Inter-American Development Bank, Washington, D.C., USA. https://unfccc.int/sites/default/files/resource/COSTOS%20CAMBIO%20CLIMATICO%281%29_Luis.pdf

Global Volcanism Program (2021) Soufriere St. Vincent. Smithsonian Institution, National Museum of Natural History, Washington, D.C., USA. https://volcano.si.edu/volcano.cfm?vn=360150

Holladay PJ, Powell RB (2013) Resident perceptions of social–ecological resilience and the sustainability of community-based tourism development in the commonwealth of Dominica. J Sustain Tour 21:1188–1211

Hynes W, Trump BD, Kirman A, Haldane A, Linkov I (2022) Systemic resilience in economics. Nat Phys 18:381–384

Keenan JM, Trump BD, Hynes W, Linkov I (2021) Exploring the convergence of resilience processes and sustainable outcomes in post-Covid, post-Glasgow economies. Sustainability 13:13415

Knutson T, Camargo SJ, Chan JC, Emanuel K et al (2020) Tropical cyclones and climate change assessment: Part II: Projected response to anthropogenic warming. Bull Am Meteor Soc 101:E303–E322

Linkov I, Trump BD, Trump J, Pescaroli G, Hynes W, Mavrodieva A, Panda A (2022a) Resilience stress testing for critical infrastructure. Int J Disaster Risk Reduct 82:103323

Linkov I, Trump BD, Trump J, Pescaroli G, Mavrodieva A, Panda A (2022b) Stress-test the resilience of critical infrastructure. Nature 603(7902):578

Linkov I, Trump BD (2019) The science and practice of resilience. Springer International, New York, NY, USA

Linkov I, Trump BD, Keisler J (2018) Risk and resilience must be independently managed. Nature 555(7694):30

Maconi D (2021) Supporting disaster preparedness in the Caribbean. Military Eng 113:66–68

Mercer J, Kelman I, Alfthan B, Kurvits T (2012) Ecosystem-based adaptation to climate change in Caribbean small island developing states: integrating local and external knowledge. Sustainability 4:1908–1932

Mohammed A, Muddeen F, Marine L, Ramlal CJ (2023) The exigency for resilient and cyber-secure critical infrastructure in the Caribbean. West Ind J Eng 45:35–49

Moulton AA, Machado MR (2019) Bouncing forward after Irma and Maria: acknowledging colonialism, problematizing resilience and thinking climate justice. J Extreme Events 6:1940003

Muñoz-Erickson TA, Meerow S, Hobbins R et al (2021) Beyond bouncing back? comparing and contesting urban resilience frames in US and Latin American contexts. Landsc Urban Plan 214:104173

Mycoo MA (2018) Beyond 1.5 C: vulnerabilities and adaptation strategies for Caribbean Small Island developing states. Reg Environ Change 18:2341–2353

Oen AMP (2019) Nature-based solutions are gaining momentum from international initiatives promoting environmental, social, and economic benefits. Integr Environ Assess Manag 15:830–831

Pagán-Trinidad I, Lopez RR, Diaz EL (2019) Education and building capacity for improving resilience of coastal infrastructure. In: Proceedings of the ASEE 126th annual conference & exposition. American Society of Engineering Education. https://monolith.asee.org/public/conferences/140/papers/27371/view

Palma-Oliveira JM, Trump BD, Wood MD, Linkov I (2018) Community-driven hypothesis testing: a solution for the tragedy of the anti-commons. Risk Anal 38:620–634
PAHO (2019) PAHO/WHO's emergency response to Hurricane Dorian. Pan-American Health Organization, Washington, D.C., USA. https://www3.paho.org/hq/index.php?option=com_content&view=article&id=15421:paho-s-emergency-response-to-hurricane-dorian&Itemid=72555&lang=en
Peterson RR (2020) Over the Caribbean top: community well-being and over-tourism in small island tourism economies. Int J Commun Well-Being
Robinson SA (2018) Adapting to climate change at the national level in Caribbean small island developing states. Island Stud J 13:79–100
Robinson SA, Butchart C (2022) Planning for climate change in Small Island Developing States: can Dominica's climate resilience and recovery Plan be a model for transformation in the Caribbean? Sustainability 14:5089
Romero-Duque LP, Trilleras JM, Castellarini F, Quijas S (2020) Ecosystem services in urban ecological infrastructure of Latin America and the Caribbean: How do they contribute to urban planning? Sci Total Environ 728:138780
Sandoval V, Sarmiento JP (2020) A neglected issue: Informal settlements, urban development, and disaster risk reduction in Latin America and the Caribbean. Disaster Prev Manag 29:731–745
Scobie M (2018) Accountability in climate change governance and Caribbean SIDS. Environ Dev Sustain 20(2):769–787
Shammin, M, Wang A, Sosland M (2022). A survey of community-based adaptation in developing countries. Climate Change Commun Resil 31
Sheller M (2021) Reconstructing tourism in the Caribbean: Connecting pandemic recovery, climate resilience and sustainable tourism through mobility justice. J Sustain Tour 29:1436–1449
Spencer N, Strobl E (2020) Hurricanes, climate change, and social welfare: evidence from the Caribbean. Clim Change 163:337–357
Steckel JC, Jakob M, Flachsland C, Kornek U, Lessmann K, Edenhofer O (2017) From climate finance toward sustainable development finance. Wiley Interdiscip Rev Clim Change 8(1):e437
Thomas A, Baptiste A, Martyr-Koller R, Pringle P, Rhiney K (2020) Climate change and small island developing states. Annu Rev Environ Resour 45:1–6
Trump BD, Poinsatte-Jones K, Elran M, Allen C, et al (2017) Social resilience and critical infrastructure systems. In: Resilience and risk: methods and application in environment, cyber and social domains. Springer, the Netherlands, pp 289–299
UNEP (2020) Small island developing states: building resilience for sustainable trade, tourism and development. United Nations environmental programme, New York, NY, USA. https://www.unep.org/news-and-stories/story/small-island-developing-states-building-resilience-sustainable-trade-tourism
UNFCC (2005) Climate change, small island developing states. Climate change secretariat, United Nations framework convention on climate change, Bonn, Germany. http://unfccc.int/resource/docs/publications/cc_sids.pdf
UN-OHRLLS (2022) About small island developing states. United Nations Office of the high representative for the least developed countries, landlocked developing countries and small island developing states, New York, NY, USA. Available on-line from: https://www.un.org/ohrlls/content/about-small-island-developing-states
UN-OHRLLS (2015) SIDS in numbers—climate change edition. United Nations Office of the High Representative for the Least Developed Countries, Landlocked Developing Countries and Small Island Developing States, New York, NY, USA. https://wedocs.unep.org/20.500.11822/7486
Vercio C, Loo LK, Green M et al (2021) Shifting focus from burnout and wellness toward individual and organizational resilience. Teach Learn Med 33:568–576
Vergara W, Rios AR, Galindo LM, Gutman P, Isbell P, Suding PH, Samaniego J (2013) The climate and development challenge for Latin America and the Caribbean: options for climate-resilient, low-carbon development. Inter-American Development Bank, Washington, D.C.,

USA. file:///C:/Users/ozner/Downloads/The-Climate-and-Development-Challenge-for-Latin-America-and-the-Caribbean-Options-for-Climate-Resilient-Low-Carbon-Development-1.pdf
Vosper E, Mitchell D, Emanuel K (2020) Extreme hurricane rainfall affecting the Caribbean, mitigated by the Paris Agreement goals. Environ Res Lett 15:104053
Wentworth M, Arnold J, Lehr M, Meadows A et al (2020) Caribbean resilience of infrastructure in the United States Virgin Islands. CEDC Summit 2020. National Science Foundation. https://par.nsf.gov/biblio/10341171-caribbean-resilience-infrastructure-united-states-virgin-islands
World Bank (2021) Acting now to protect the human capital of our children: The costs of and response to COVID-19 pandemic's impact on the education sector in Latin America and the Caribbean. World Bank, Washington, D.C., USA. https://openknowledge.worldbank.org/entities/publication/01e12f4e-e6a2-50fd-b929-02d5e4307c6f
Zhang H, Dolan C, Jing SM, Uyimleshi J, Dodd P (2019) Bounce forward: economic recovery in post-disaster Fukushima. Sustainability 11:6736

Index

C. D. Metcalfe and E. R. Bennett (eds.), *Building Resilience to Climate Change in Small Island Developing States (SIDS) in the Caribbean*, Environmental Contamination Remediation and Management, https://doi.org/10.1007/978-3-031-37376-3

GPSR Compliance

The European Union's (EU) General Product Safety Regulation (GPSR) is a set of rules that requires consumer products to be safe and our obligations to ensure this.

If you have any concerns about our products, you can contact us on ProductSafety@springernature.com

In case Publisher is established outside the EU, the EU authorized representative is:

Springer Nature Customer Service Center GmbH
Europaplatz 3
69115 Heidelberg, Germany

Batch number: 10726062

Printed by Printforce, the Netherlands